Hydrologic
and Hydraulic

Modeling Support

with
Geographic
Information
Systems

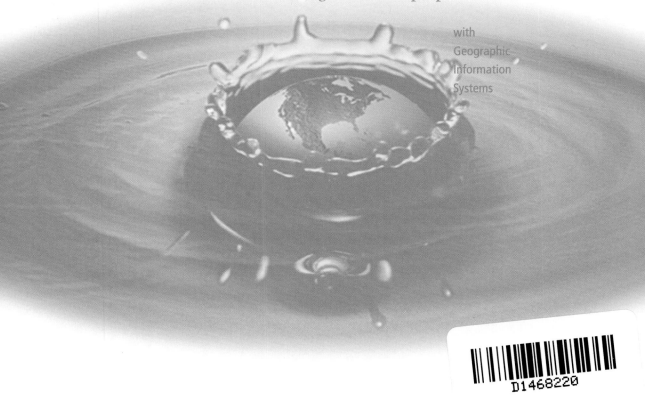

Compiled and edited by Dr. David Maidment and Dr. Dean Djokic

ESRI PRESS

Environmental Systems Research Institute, Inc.
  Hydrologic and Hydraulic Modeling Support with Geographic Information Systems
  ISBN 1-879102-80-3

First printing: April 2000

10 9 8 7 6 5 4 3 2 1

Printed in the United States of America.

Published by Environmental Systems Research Institute, Inc., 380 New York Street, Redlands, California 92373-8100.

ESRI PRESS books are available to resellers worldwide through Independent Publishers Group (IPG). For information on volume discounts, or to place an order, call IPG at 1-800-888-4741 in the United States, or at 312-337-0747 outside the United States.

Contents

Introduction

BACKGROUND THIS BOOK is a compilation of the invited papers in the Water Resources track at the 1999 ESRI® International User Conference held in San Diego from July 26 through July 30, 1999. The idea behind the invited papers and sessions was to produce a high-quality forum for dealing with specific GIS issues in water resources. It was intended that the results of these sessions would define the state of the art for that particular topic and serve as a reference for the larger community interested in GIS applications in water resources.

The specific topics presented in this book deal with terrain modeling and GIS support for hydrologic and hydraulic modeling, with a focus on NexGen models developed by the Hydrologic Engineering Center (HEC) of the U.S. Army Corps of Engineers. Although the focus is on HEC model support, the GIS principles presented here apply to most water resources models.

The technology and methods for GIS support of water resources applications (with a focus on water resources modeling) have been available for many years as indicated by the extensive bibliography in the field. There are several reasons why the available technology has not been more widely used in day-to-day water resources operations:

- Suitable data has been lacking.

- The expense of GIS technology has limited its use to larger organizations, while most of the services in the field are in the domain of small consulting companies.

- The engineering community has not been educated enough in GIS, while the GIS community has not being educated enough in engineering fields, making cross-discipline communication and implementation difficult.

These barriers are being brought down quickly. The development of national hydrographic and elevation data sets, and new techniques for elevation derivation that are feasible and accurate, make core data much easier to obtain. Advanced desktop GIS software and the lower cost of required hardware make GIS functionality available to smaller organizations. Finally, the development of well-established data-processing procedures and packaged GIS solutions for specific water resources tasks (including education and training) make GIS implementation much simpler and less threatening to traditionally non-GIS organizations.

The invited papers and subsequently this book have a role in this confluence of events. The intent is to provide in one place the overview of the GIS technology, methods, issues, and tools that are established and can be used by water resources practitioners to put GIS technology to quick, practical, and efficient use in the area of hydrologic and hydraulic modeling. It also provides to the GIS community an overview of GIS applications and issues that the water resources community is interested in.

CONTENT

The book contains three main sections. The papers within each section are designed so that the first presents the basic principles, the next describes available tools and techniques, and the final paper(s) presents practical application of the GIS tools and techniques to real-world modeling projects.

The first section deals with the use of digital elevation models (DEMs) in water resources modeling. It provides the foundation for understanding the role and impact of DEMs in modeling support. In "Digital Elevation Model Issues in Water Resources Modeling," Garbrecht and Martz present capabilities and limitations of DEMs in water resources modeling support. They address issues of data availability, quality, and resolution, as well as capabilities and limitations in automated extraction of topographic parameters from DEMs.

In "Preparation of DEMs for Use in Environmental Modeling Analysis," Saunders discusses techniques for manipulating DEMs by integrating vector hydrography to obtain the hydrologically correct terrain representation and best results for DEM-derived output. Perez, in "Source Water Protection Project: A Comparison of Watershed Delineation Methods in ARC/INFO and ArcView GIS," presents the efforts by the Washington State DOT to use DEM and GIS techniques to delineate the Source Water Protection Areas (contributing watershed area) for all surface water intakes statewide.

The second section presents the issues of rainfall-runoff modeling using GIS and HEC-HMS. The Hydrologic Modeling System (HMS), developed by the Hydrologic Engineering Center (HEC) of the U.S. Army Corps of Engineers, is one of the most popular computer models for simulation of precipitation-runoff processes. "DEM Preprocessing for Efficient Watershed Delineation" by Djokic and Ye presents a methodology for DEM preprocessing that facilitates efficient and interactive watershed delineation. This methodology is used as a basis for interactive basin development in GIS tools for HMS model support.

In "GIS Tools for HMS Modeling Support," Olivera and Maidment present a set of raster- and vector-based GIS tools for determination of hydrologic elements and their connectivity, calculation of their parameters, and preparation of an input file for HMS. These tools allow automated model setup that is significantly faster than traditional methods and leads to reproducible results. In "Hydrologic Model of the Buffalo Bayou Using GIS," Doan applies the previously described tools to a watershed that covers most of the Houston metropolitan area. The analysis is based on USGS DEMs at 30-meter cell resolution and stream data from USGS digital line graphs (DLGs) and EPA river reach files (RF1). Physical watershed parameters were extracted using GIS to support hydrologic parameters computation. The model uses grid-based radar (NEXRAD) rainfall in a Standard Hydrologic Grid.

The third and final section of the book deals with the issues related to floodplain modeling using GIS and HEC-RAS. The River Analysis System (RAS), developed by the HEC, is the most popular computer model for simulation of one-dimensional, gradually varied flow in open channels. This section of the book involves a switch from the hydrologic, watershed-based processes and their modeling based on DEM terrain representation described in the previous sections to hydraulic, stream-based processes and their modeling based on triangular irregular network (TIN) terrain representation.

Long, in "Development of Digital Terrain Representation for Use in River Modeling," presents techniques for development of quality TINs that can support RAS modeling. The issues of underwater terrain modeling and inclusion of old cross-section (HEC-2) data are discussed in particular. In "HEC-GeoRAS: Linking GIS to Hydraulic Analysis Using ARC/INFO and HEC-RAS," Ackerman, Evans, and

Brunner present GIS tools developed at HEC to support RAS modeling. These tools speed up spatial data pre- and postprocessing and allow the engineer to concentrate on hydraulic principles during model development and analysis, rather than focusing on applying the GIS.

In "Floodplain Determination Using ArcView GIS and HEC-RAS," Kraus presents the application of GIS tools for RAS modeling support on a floodplain study for the National Flood Insurance Program. Cross sections were located to coincide with field-surveyed cross sections of the channels. The extracted data was imported into HEC-RAS and combined with the field-surveyed data to construct full floodplain cross sections. HEC-RAS-computed water surface elevations along the channels were transferred to ArcView® GIS where the floodplain limits were automatically determined. Dodson and Li, in "The Accuracy and Efficiency of GIS-Based Floodplain Determinations," compare the accuracy and efficiency of computing floodplain elevations and mapping floodplain boundaries using the traditional and the GIS approach.

This book presents the most up-to-date survey presently available of the application of GIS to flood hydrology and hydraulics. Its primary focus is on the use of grids or triangulated irregular networks to describe the watershed and stream channel terrain surfaces. It demonstrates that the techniques in this field have evolved into a sound, reliable technological base, which can be used with confidence to construct hydrologic and hydraulic models based on GIS data.

What further innovations can be expected in this field in the near future? One of the key GIS data structures having potential for application in the water resources field is the network model. The river network is the backbone structure for describing the motion of water through the landscape. It connects watersheds with their stream channels, describes the connectivity of points along rivers, and determines the ordering of flow as it passes from one river reach to the next. Up to this point, the GIS network model has been mainly applied to transportation problems involving routing vehicles from one location to another on a street system.

With the advent of the new geodatabase concepts in ArcInfo™ 8, which will also later be incorporated into ArcView GIS, the network model will be able to be incorporated much more directly into water resource modeling than has been the case. This innovation in software is enhanced by the fact that the National Hydrography

Dataset has recently been released by the USGS and EPA, providing a comprehensive description at 1:100,000 scale of the river network of the United States. The combination of raster data, grids, and TINs, as described in this book, along with a stronger network approach, will enhance the capability to connect multiple hydrologic and hydraulic models to a common geodatabase of a region.

The design of a geodatabase for water resource applications, called the Arc Hydrology Data Model, is presently being carried out by a GIS in Water Resources Consortium formed by the Center for Research in Water Resources (CRWR) of the University of Texas at Austin, by ESRI, and by representatives of government agencies, consulting firms, and academic institutions. Further information about this effort can be found at www.crwr.utexas.edu/giswr. The goals of this effort are to support mapping of water features, linear referencing of locations along a river network, and dynamic modeling of water resources. The core Arc Hydrology Data Model will provide a foundation for extension and customization to fit the needs of particular organizations and physical settings. This data model is being built using the object representation structure of ArcInfo 8, which will yield new opportunities for the closer synthesis of GIS and hydrology.

While the new geodatabase concepts will take time to develop, the key ideas presented in this book will remain valid and will be incorporated in the new software systems. The editors of this book have been involved in GIS hydrology for more than a decade. From this perspective, we have seen the growth of this field from a few individuals working in research institutions to a mature community of GIS and hydrology professionals dedicated to the common goal of putting hydrologic analysis on a more secure footing of geospatial data and software engineering techniques. Progress in this field depends on many factors, including data development, software innovations, research insights, educational programs, and most of all, the dedication of the people involved in the GIS hydrology community. The publication of this book is a milestone along the journey of development of this field. We hope that you will find it useful and informative.

Dean Djokic

David Maidment

Paper 1 **Digital Elevation Model Issues in Water Resources Modeling**

JURGEN GARBRECHT AND LAWRENCE W. MARTZ

GRAZINGLANDS RESEARCH LABORATORY

USDA–ARS

EL RENO, OKLAHOMA

ABSTRACT

TOPOGRAPHY plays an important role in the distribution and flux of water and energy within natural landscapes. The automated extraction of topographic parameters from DEMs is recognized as a viable alternative to traditional surveys and manual evaluation of topographic maps, particularly as the quality and coverage of DEM data increase. The capabilities and limitations of DEMs for use in water resources model applications are reviewed. Specifically, data availability, quality, and resolution are discussed from an application perspective. Issues related to the automated extraction of topographic and drainage information from DEMs are presented. These include the identification of drainage in the presence of pits and flat areas in the DEM, the determining role of channel source definition in drainage network configuration, and network analysis capabilities for raster networks. Also presented are the latest research results regarding the reduction of distributed subcatchment properties into a representative value for the subcatchments. Increasing quality and resolution of DEM products, new raster-processing methodologies, and expanding GIS capabilities and linkages with water resources models are expected to lead to a heavier reliance on DEMs as a source of topographic and surface drainage information.

INTRODUCTION

Hydrologic process and water resource issues are commonly investigated by use of distributed watershed models. These watershed models require physiographic information such as configuration of the channel network, location of drainage divides, channel length and slope, and subcatchment geometric properties. Traditionally, these parameters are obtained from maps or field surveys. Over the last two decades this information has been increasingly derived directly from digital representations of the topography (Jenson and Domingue, 1988; Mark, 1984; Moore et al., 1991; Martz and Garbrecht, 1992). The digital representation of the topography is called a digital elevation model (DEM). The automated derivation of topographic watershed data from DEMs is faster, less subjective, and provides more reproducible measurements than traditional manual techniques applied to topographic maps (Tribe, 1992). Digital data generated by this approach also have the advantage of being readily imported and analyzed by geographic information systems (GIS). The technological advances provided by GIS and the increasing availability and quality of DEMs have greatly expanded the application potential of DEMs to many hydrologic, hydraulic, water resources, and environmental investigations (Moore et al., 1991). In this paper the production, availability, quality, resolution, and capabilities of DEMs are reviewed and discussed with respect to the derivation of topographic data in support of hydrologic and water resources investigations. This paper covers DEMs of natural landscapes only and does not extend to urbanized settings where small-scale and manmade structures such as street gutters, inlets, drainage ditches, and culverts control surface drainage patterns.

DEM PRODUCTION, QUALITY, AND AVAILABILITY

The most common DEM data structure is the raster or grid structure. This normally consists of a matrix of square grid cells with the mean cell elevation stored in a two-dimensional array. Location of a cell in geographic space is implicit from the row and column location of the cell within the array, provided that the boundary coordinates (georeferences) of the array are known. Grid DEMs are widely available and used because of their simplicity, processing ease, and computational efficiency (Martz and Garbrecht, 1992). Disadvantages include grid size dependency of certain computed topographic parameters (Fairfield and Leymarie, 1991) and inability to locally adjust the grid size to the dimensions of topographic land surface features. Other DEM data structures, such as the triangulated irregular network and contour-based structures, have overcome some of the disadvantages of grid DEMs. However, they have shortcomings of their own and are not as widely available and used as grid DEMs. The remainder of this paper will focus on the popular grid-type DEMs.

In the United States, the most widely available DEMs are those distributed by the U.S. Geological Survey (USGS). They are produced using elevation data derived from existing contour maps, digitized elevations, and photogrammetric stereomodels based on aerial photographs and satellite remote-sensing images. The USGS 7.5-minute DEMs have a grid spacing of 30 by 30 meters and are based on the Universal Transverse Mercator (UTM) georeferencing system. These DEMs provide coverage in 7.5-by-7.5-minute blocks, and each block provides the same coverage as a standard USGS 7.5-minute map series quadrangle (USGS, 1990). The USGS 1-degree DEMs have a grid spacing of 3 by 3 arc-seconds and provide coverage in 1-by-1-degree blocks. Two coverages provide the same coverage as a standard USGS 1-by-2-degree map series quadrangle. The USGS 30-minute DEMs have a grid spacing of 2 by 2 arc-seconds and consist of four 15-by-15-minute DEM blocks. Two 30-minute DEMs provide the same coverage as a standard USGS 30-by-60-minute map series quadrangle. All USGS DEMs provide elevation values in integer feet or meters.

DEMs produced by the USGS are classified into three levels of increasing quality. Level 1 classification is generally reserved for data derived from scanning National High-Altitude Photography Program, National Aerial Photography Program, or equivalent photography. A vertical Root Mean Square Error (RMSE) of 7 meters is the targeted accuracy standard, and a RMSE of 15 meters is the maximum permitted. Level 2 classification is for elevation data sets that have been processed or smoothed for consistency and edited to remove identifiable systematic errors. A RMSE of one-half of the original map contour interval is the maximum permitted. There are no errors greater than one contour interval in magnitude. Level 3 classification DEMs are derived from digital line graph (DLG) data by using selected elements from both hypsography (contours, spot elevations) and hydrography (lakes, shorelines, drainage). If necessary, ridgelines and major transportation features are also included in the derivation. A RMSE of one-third of the contour interval is the maximum permitted. There are no errors greater than two-thirds of the contour interval in magnitude. Most data produced within the last decade fall into the level 2 classification. The availability of level 3 DEMs is very limited.

The USGS, Earth Science Information Center, Reston, Virginia, offers a variety of digital elevation data products (USGS, 1990). Other sources for DEM data include the former Defense Mapping Agency (DMA) (now the National Imagery and Mapping Agency, NIMA), and the National Geophysical Data Center (NGDC) of the National Oceanic and Atmospheric Administration (NOAA). Custom DEM data can also be obtained through a number of commercial providers. New technologies, such as Laser Altimetry (LA) (Ritchie, 1995) and Radar Interferometry (RI) (Zebker and Goldstein, 1986), are currently being explored for production of high-quality and high-resolution global DEMs (Gesch, 1994).

DEM ACCURACY CONSIDERATIONS IN WATER RESOURCE APPLICATIONS

DEMs are used in water resources projects to identify drainage features such as ridges, valley bottoms, channel networks, and surface drainage patterns, and to quantify subcatchment and channel properties such as size, length, and slope. The accuracy of this topographic information is a function both of the quality and resolution of the DEM, and of the DEM processing algorithms used to extract this information.

The suitability of a USGS DEM for water resources projects depends largely on the DEM production techniques. USGS 7.5-minute DEMs produced before 1988 were mainly based on manual profiling of photogrammetric stereomodels (USGS, 1990). In low-relief landscapes, the resulting DEMs often display systematic east–west striping patterns that can make them unsuitable for parameterization of drainage features (Garbrecht and Starks, 1995). Figure 1a shows two adjacent DEMs that were produced by different techniques. The left side illustrates the east–west striping pattern associated with the manual profiling of photogrammetric stereomodels.

The impact of the striping on drainage studies is threefold. First, the outlines of drainage features, such as depressions or drainage paths, are not well defined. Boundaries of shallow features that have a north-to-south orientation (i.e., perpendicular to the striping) are often represented in the DEM as ragged lines having east-to-west indentations. Second, drainage paths are systematically biased in the

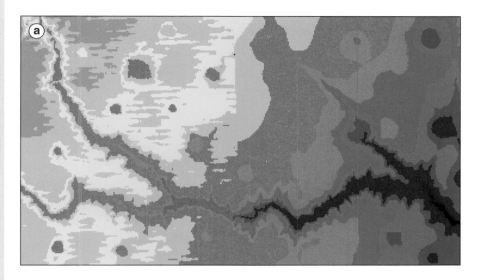

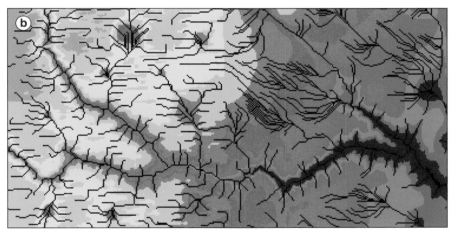

Figure 1. Coverage of two adjacent USGS 7.5-minute DEMs near Amarillo, Texas; (a) DEM elevation values; (b) GIS-derived drainage network.

east-to-west direction because of flow draining into and following the artificial elevation stripes. Figure 1b illustrates the differences in drainage pattern as a result of the striping pattern in the elevation data. And, third, the striping may introduce drainage blockages in the north-to-south flow component. These drainage blockages can produce artificial depressions of varying sizes. The source of the striping is a combination of human and algorithmic errors associated with the manual profiling method (B. Kunert, USGS, Rolla, Mid Continent Mapping Center, personal communication). While these "striping" errors are well recognized (Garbrecht and Starks, 1995), they are within the accuracy standards of the USGS (1990). While most DEMs being developed today are derived from DLGs and are processed to level 2 standards, many DEMs being distributed today were developed in the past and meet only level 1 standards.

Level 1 standards demand a RMSE value of 7 meters, with a maximum permitted value of 15 meters. An absolute elevation error tolerance of 50 meters is set for blunders for any grid node when compared to the true height from mean sea level. Also, any array of 49 contiguous elevation points shall not be in error by more than 21 meters (USGS, 1990). These tolerances in elevation are large for drainage investigations since an elevation difference of 1 or 2 meters can affect flow path and runoff characteristics.

DEM horizontal resolution and its ratio to vertical resolution can have a significant bearing on computed land surface parameters that involve differences in elevations. For example, slope is computed as the difference in elevation between two adjacent pixels divided by the distance between them. Since DEM elevations are generally reported in full meters or feet, the computed slope can only take on a limited number of discreet values. For a 30-meter DEM with elevations reported in meters, a slope value between two pixels can be zero (no change in elevation), 0.033 (1-meter change in elevation), or a multiple thereof. Such increments may be adequate to represent slope values in mountainous terrain, but for flat areas, such as the Great Plains of the United States, a 1-meter vertical DEM resolution is insufficient to provide accurate local slope values. Thus, DEMs of low-relief landscape and limited vertical resolution do not lend themselves well to an accurate determination of drainage slopes and precise location of channels and ridges.

The problems of DEM quality and resolution can generally not be overcome by smoothing or averaging the DEM. Such approaches simply cover up the problems without increasing the quality of the output. The easiest solution to overcome the described resolution problems is to custom produce a DEM with a prespecified horizontal-to-vertical resolution ratio, or to use a high-resolution DEM produced by more advanced methods. Other solutions include the use of DEM analysis methods that are designed to overcome problems associated with digital representations of low-relief landscapes by DEMs of limited resolution (Garbrecht and Martz, 1999a). Examples of such problems are the increased occurrence and size of flat areas and spurious pits. Pits are cells that have no adjacent cell at a lower elevation and, consequently, have no downslope flow path to an adjacent cell. On the other hand, flat areas are characterized by adjacent cells with the same elevation values. Pits and flat areas occur in most raster DEMs, but are predominant in limited-resolution DEMs of low-relief landscapes. Figure 2 shows the spatial distribution and extent of pits and flat areas (areas in green) of a DEM of a watershed in central Oklahoma. The predominance of pits and flat areas in the valley bottoms (low-relief areas) are clearly visible. Pits are usually viewed as spurious features that arise from interpolation errors during DEM generation and truncation of interpolated values on output (O'Callaghan and Mark, 1984; Mark, 1988; Fairfield and Leymarie, 1991). Pits are a major difficulty for DEM evaluation methods that rely on the overland flow simulation approach to drainage analysis because a lack of downslope flow paths leads to incomplete drainage pattern definition. The drainage identification problems for flat areas are similar to those encountered for pits. This subject is addressed in greater depth in the section on automated extraction of drainage features from DEMs.

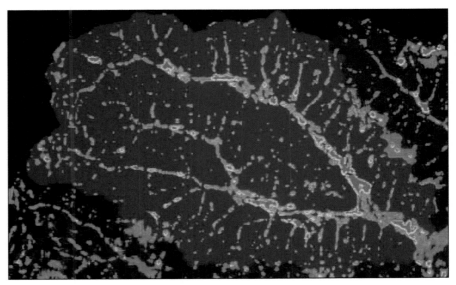

Figure 2. Spatial distribution and extent of pits and flat areas in a DEM of a watershed in central Oklahoma.

**DEM SELECTION
CRITERIA**

Both quality and resolution must be considered in selecting a DEM for hydrologic modeling. Quality refers to the accuracy of the elevation data, while resolution refers to the precision of the data, specifically to the horizontal grid spacing and vertical elevation incrementation. Quality and resolution must be consistent with the scale and model of the physical process under consideration and with the study objectives. For many applications of physical-process-based environmental models, the USGS 30-by-30-meter DEM data (levels 1 and 2) has broad accuracy standards and a rather coarse resolution with documented shortcomings (Garbrecht and Starks, 1995; Ostman, 1987; Topographic Science Working Group, 1988). In particular, surface drainage identification is difficult in low-relief landscapes, as is derivation of related information such as slope and landform curvature. Research is underway to assess the impact of accuracy limitations, noise, and low resolution of DEM data on modeling results. Examples of such studies include Wolock and Price (1994) and Zhang and Montgomery (1994).

The accuracy of drainage features extracted from DEMs as a function of DEM resolution was investigated by Garbrecht and Martz (1994). The horizontal resolution of a DEM with an original grid spacing of 30 meters was decreased by cell aggregation. Selected drainage features for several hypothetical channel network configurations were extracted for a range of DEM resolutions using the TOPAZ software (Garbrecht and Martz, 1994). Figure 3 illustrates the loss of accuracy from left to right with increasing grid coefficient. The grid coefficient is the area of a cell divided by the network reference area, which is the mean subcatchment area. The values shown are for selected network features such as channel source area, number of channels, channel length, and drainage density. The sensitivity analysis suggested that a DEM should have a grid area of less than 5 percent of the network reference area to reproduce the selected drainage features with an accuracy of about 10 percent. It was concluded that the grid resolution dependency was introduced by the inability of a DEM to accurately reproduce drainage features that are at the same scale as the spatial resolution of the DEM. For sinuous channels, this results in shorter channel lengths, and for networks with high drainage density, it leads to channel and drainage area capturing. Channel and drainage area capturing occurs when the DEM resolution can no longer resolve the separation between channels or drainage boundaries. In such situations, the number of channels, the size of direct drainage areas, and the channel network pattern may depart considerably from the one obtained by a high-resolution DEM. Thus, if small drainage features are important, the resolution must be selected relative to the size of these features.

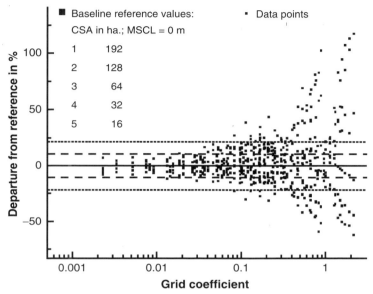

Figure 3. Departures from reference for selected drainage features versus grid coefficients (from Garbrecht and Martz, 1994).

Zhang and Montgomery (1994) used high-resolution digital elevation data from two small catchments in the western United States to examine the effect of DEM resolution on the portrayal of the land surface and results of hydrologic simulations. They evaluated DEMs that had resolutions of 2 through 90 meters and concluded that grid sizes of 10 meters would suffice for many DEM-based applications of geomorphic and hydrologic modeling. Commenting on this research, Garbrecht and Martz (1996) suggested that the selection of DEM resolution for simulation applications depends not only on the scale of the processes being modeled, but also on the numerical simulation approach and the specific landscape parameters to be extracted from the DEM. For hydrologic models that operate on a grid approach, landscape parameters and simulation processes are determined at every grid cell. Hence, the data volume and computational resources are proportional to the number of grid cells which themselves increase quadratically for each doubling of the horizontal DEM resolution. This can be a limiting factor for practical applications and encourages the selection of lower resolution DEMs. On the other hand, models that use subcatchments as their operating unit depend only on the DEM to extract representative parameters for the entire subcatchment, and the processes are simulated at the subcatchment level, independently of DEM resolution. This approach is less demanding with regard to data volume and computational resources. It often is also less sensitive to DEM noise and resolution because representative landscape parameters for subcatchments, such as hillslope length, width, and slope, are derived from many grid cells. As a result, the effects of grid-induced local variability and discrete incrementation are largely eliminated by the averaging process over the subcatchment.

In a study by Seybert (1996), GIS coverages of land use, soils, and elevation (DEM) were used to study the effect of spatial data resolution degradation on the output of an event-based surface runoff model. The study was performed on a 7.27-square-kilometer (2.81-square-mile) agricultural watershed in central Pennsylvania. Square cell dimensions ranging from 5 meters to 500 meters were investigated, beginning with spatial data layers of the finest resolution and systematically degrading the data to the coarsest resolution. Results of the spatial data resolution study indicate that volume estimates in the model are less sensitive to spatial resolution change than peak flow estimates. Also, increasing the number of subcatchments in the watershed representation caused the model to increase estimates of runoff volume and peak flow. The ratio of average subcatchment area to grid cell area was used as an indicator of spatial resolution, and an overall catchment-to-grid ratio of about 102 was found to be an acceptable threshold of spatial resolution for reasonable model results.

The above discussion provides some broad indications as to appropriate DEM resolutions for specific applications, but there appears to be no guidelines for DEM resolution for general applications. In theory, the DEM resolution should be selected as a function of the size of the land surface features that are to be resolved, the scale of the processes that are modeled, and the numerical model used to model the processes. In practice, however, the selection of DEM resolution for a particular application is often driven by data availability, judgment, test applications, experience, and, last but not least, cost.

AUTOMATED EXTRACTION OF DRAINAGE FEATURES

The major issues with the derivation of surface drainage, drainage networks, and associated topologic information from DEMs are related to the resolution and quality of the DEM and to the methodology used to derive this information. Issues of DEM resolution and quality have been addressed in an earlier section. This section addresses the raster-processing methodology used in the derivation of the topographic/topologic drainage information from USGS DEMs. Discussions are limited to methodologies based on the D-8 method (described below). The D-8 method is selected here because it is a simple and widely used raster DEM processing method.

The D-8 method (Fairfield and Leymarie, 1991) defines the drainage network from raster DEMs using an overland flow analog. The method identifies the steepest downslope flow path between each cell of a raster DEM and its eight neighbors (hence the method's name), and defines this path as the only flow path leaving the raster cell. The drainage network is identified by selecting a threshold catchment area at the bottom of which a source channel originates; all cells with a catchment area greater than this threshold catchment area are classified as part of the drainage network. This drainage network identification approach is simple and directly generates connected networks (Martz and Garbrecht, 1992). The use of the D-8 method for catchment area and drainage network analysis recently has been criticized on the grounds that it permits flow only in one direction, away from a DEM cell. This fails to represent adequately divergent flow over convex slopes (Freeman, 1991; Quinn et al., 1991; Costa-Cabral and Burges, 1994) and can lead to a bias in flow path orientation (Fairfield and Leymarie, 1991). Although the multiple-flow-direction algorithm seems to give superior results in the headwater region of a source channel, a single-flow-direction algorithm is superior in zones of convergent flow and along well-defined valleys (Freeman, 1991; Quinn et al., 1991). Thus, for overland flow analysis on hillslopes, the multiple-flow approach may be more appropriate. However, if the primary objective is the delineation of the drainage network for large drainage areas with well-developed channels, the use of a single-flow-direction algorithm seems more appropriate (Martz and Garbrecht, 1992).

The D-8 method, as well as many other approaches, has difficulties identifying surface drainage in the presence of depressions, flat areas, and flow blockages (Garbrecht and Starks, 1995; Martz and Garbrecht, 1998). These features are often the result of data noise, interpolation errors, and systematic production errors in DEM elevation values. Such features occur in most DEMs and are viewed as spurious, mainly because of their predominantly numerical origin. The difficulties arise from the fact that raster cells in depressions, sinks, and flat areas have no neighboring cells at a lower elevation, and, consequently, have no downslope flow path to a neighbor cell. It is common practice to remove the depressions and flat areas prior to drainage identification.

SURFACE DRAINAGE IDENTIFICATION FROM RASTER DEMS

A number of methods have been developed for treating sinks and flat areas in DEMs for automated analysis of hydrographic terrain features. Band (1986) simply increases the elevation of sink cells until a downslope path to a cell becomes available, under the constraint that flow may not return to a sink cell. O'Callaghan and Mark (1984) suggest smoothing a DEM prior to analysis to reduce the size and number of sinks. Both Jenson and Domingue (1988) and Martz and De Jong (1988) provide methods for treating sinks in a more general and effective manner (Freeman, 1991). These methods cope with complex topographic situations such as nested depressions, depressions with flat areas, and truncated depressions and flat areas at the edge of the DEM. They involve "filling" each depression in the DEM to the elevation of the lowest overflow point out of the sink. This approach implies that all sinks are the result of underestimation of elevation (Martz and Garbrecht, 1998). However, some sinks arise from obstruction of flow paths by overestimated elevations. In such cases, breaching the obstruction is more appropriate than filling the sink created by the obstruction (Martz and Garbrecht, 1999a). Obstruction breaching is particularly effective in DEMs of landscapes that have a low relief relative to the vertical resolution of the DEM because sinks caused by flow path obstruction are more prevalent in these situations. A combined filling and breaching method was presented by Garbrecht et al. (1996).

Once the sinks in a DEM are removed by breaching and filling, the resulting flat surface must still be interpreted to define the surface drainage pattern. Flat surfaces do not only result from sink filling, but can also result from too low a vertical and/or horizontal DEM resolution to adequately represent the landscape, or from a truly flat landscape (which seldom occurs). Several approaches for defining surface drainage across flat areas have been discussed earlier (Band,

1986; Jenson and Domingue, 1988; Martz and De Jong, 1988). Other methods rely on techniques ranging from landscape smoothing to arbitrary flow direction assignment. A discussion of various approaches can be found in Tribe (1992) and subsequent comments by Martz and Garbrecht (1995). A more recent approach has been presented by Garbrecht and Martz (1995). This approach recognizes that surface drainage in natural landscapes is toward lower terrain and away from higher terrain. To reproduce such a trend on a flat surface, two shallow gradients are imposed on flat surfaces to force flow away from higher terrain surrounding the flat surface and attracting flow toward lower terrain on the edge of the flat surface. This approach results in a convergent flow direction pattern over the flat surface that is also consistent with the topography surrounding the flat surface (Garbrecht et al., 1996). Figure 4a shows the drainage pattern of a hypothetical mountain saddle which consists of a flat surface at the center between higher terrain to the right and left, and three locations of lower terrain, one at the top and two at the bottom. The computed drainage, as well as flow convergence and consistency with the surrounding terrain configuration, is a great improvement over drainage patterns from methods that are plagued by the parallel flow problem (figure 4b). It must, however, be recognized that any method of drainage identification in flat areas is an approximation and may not accurately reflect the actual drainage pattern which may follow channel incisions that are too small to be resolved at the resolution of the digital landscape.

DRAINAGE NETWORK EXTRACTION

A drainage network can be extracted from a DEM with an arbitrary drainage density or resolution (Tarboton et al., 1991). The characteristics of the extracted network depend on the definition of channel sources on the digital land surface topography. Once the channel sources are defined, the essential topology and morphometric characteristics of the drainage network are implicitly defined because of their close dependence on channel source definition. Thus, the proper identification of channel sources is critical for extracting a representative drainage network from DEMs.

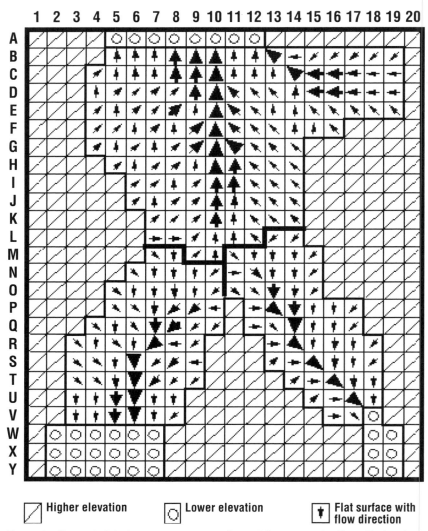

Figure 4a. Computed drainage pattern over a flat saddle topography (from Garbrecht and Martz, 1997b). Arrows indicate drainage direction; arrow size is representative of upstream drainage area; drainage divides are indicated by heavy lines; drainage computed after Garbrecht and Martz (1996).

The fundamental concepts that deal with channel initiation can readily be found in the scientific literature (Montgomery and Dietrich, 1988, 1989, and 1992). Two prevailing methods for network sources in DEMs are the constant threshold area and the slope-dependent critical support area method (Montgomery and Foufoula-Georgiou, 1993; Tribe, 1992). The constant threshold area method assumes

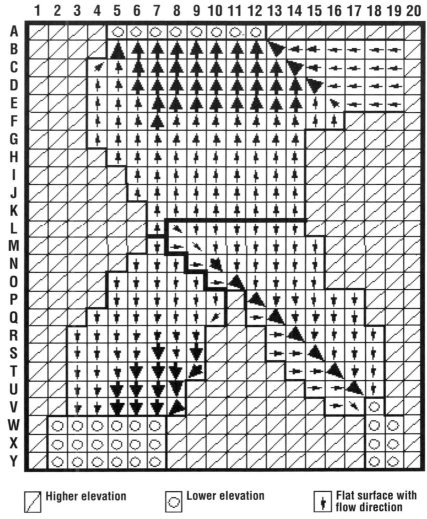

Higher elevation   Lower elevation   Flat surface with flow direction

Figure 4b. Computed drainage pattern over a flat saddle topography (from Garbrecht and Martz, 1997b). Arrows indicate drainage direction; arrow size is representative of upstream drainage area; drainage divides are indicated by heavy lines; drainage computed after Martz and De Jong (1988) and Jenson and Domingue (1988).

that channel sources represent the transition between the convex profile of the hillslope (sheet flow dominated) and the concave profiles of the channel slope (channel discharge dominated). The constant threshold area method has found widespread application (Band, 1986; Morris and Heerdegen, 1988; Tarboton et al., 1991; Gardner et al., 1991). The application details, use, and implications of this method are discussed by Tarboton et al. (1991) and Montgomery and Foufoula-Georgiou (1993). Garbrecht and Martz (1995) broadened the use of the constant threshold area method by allowing the threshold area to vary within the DEM. This is particularly useful in large watersheds in which geology and drainage network characteristics display distinct spatial patterns. The user must, however, establish *a priori* the values of the constant threshold areas and their region of application.

The slope-dependent critical support area method assumes that the channel source represents an erosion threshold. This assumption implies that the channel source is the result of a change in sediment transport processes from sheet flow to concentrated flow, rather than a spatial transition in longitudinal slope profiles. This method was presented by Montgomery and Foufoula-Georgiou (1993) and is based on the channel initiation work by Dietrich et al. (1993). The main differences between the networks defined by the constant and slope-dependent threshold support area methods are the spatial variability of the slope. Montgomery and Foufoula-Georgiou (1993) report that with the slope-dependent threshold method, drainage density is greater in steeper portions of the catchment, as is found in natural landscapes.

The constant threshold area method appears more practical in application, as is apparent by its widespread use. The current preference for this method can be attributed to the fact that local slope values are difficult to obtain from DEMs. Indeed, a DEM with horizontal and vertical resolution of 30 meters and 1 meter, respectively, can only produce local slopes of zero, 0.03, or increments thereof. Thus, an accurate estimation of local slope requires either a high-resolution DEM or field measurements.

Another issue with drainage networks extracted from DEMs is the precise positioning of channels in the digital landscape. Comparisons with actual maps or aerial photos often show discrepancies, particularly in low-relief landscapes. The primary reason for this discrepancy is that digital landscapes cannot capture important topographic information below the DEM resolution. Though the channel position in the digital landscape is consistent with the digital topography, it may not reflect the actual drainage path in the field. From a practical point of view, this dilemma can be overcome by "burning in" the path of the channels along predigitized pathways. This is achieved by artificially lowering the elevation of the DEM cells along digitized lines or raising the entire DEM except along stream paths (Cluis et al., 1996; Maidment et al., 1996). However, caution is advised with this method because it may produce flow paths that are not consistent with the digital topography.

DRAINAGE NETWORK TOPOLOGY

Once a channel network is extracted from a DEM, it is usually displayed as a string of raster cells. For these images to be useful in hydrologic and runoff modeling, individual channel links and adjacent contributing areas must be explicitly identified and associated with topologic information for upstream and downstream connections. Such channel indexing is straightforward in vector GIS, but usually not easy when working with raster data. Yet, the spatial organization and connectivity of the channel network is fundamental for flow routing and automated linkage of raster GIS information of the network and traditional surface runoff modeling. Garbrecht and Martz (1997a) have proposed an algorithm that interprets a GIS raster image of a network, indexes the channel links and network nodes, and organizes the channels into a sequence for cascade flow routing (Garbrecht, 1988). The algorithm uses a cell-by-cell trace along the channels of the network to identify for each channel link the Strahler order, the index, the sequence for cascade flow routing, the upstream inflow tributaries, the downstream connection, and the channel longitudinal slope and length. This data is provided in tabular format and can be used as input data for hydrologic modeling.

CALCULATION OF DISTRIBUTED SUBCATCHMENT PROPERTIES

The traditional automated extraction of drainage features from DEMs has focused on the identification of drainage boundaries, channels, and watershed segmentation (Band, 1986; Jenson and Domingue, 1988; Martz and Garbrecht, 1992; Wolock and McCabe, 1995; Tarboton et al., 1991). The identification of distributed properties for subcatchments from digital elevation models is a relatively recent area of exploration. Contrary to the uniquely identifiable topographic features such as drainage divides and channels, distributed properties require a method or model to reduce the distributed information into a representative value for the entire subcatchment. Thus, a representative subcatchment property can have different values depending on the underlying model used for data reduction. The discrepancies between the different values are not approximation errors, but result from the different interpretations of the subcatchment property by the data reduction models. These differences can have significant implications for distributed rainfall-runoff and erosion modeling. In the following, alternative models for subcatchment length and slope calculations, as well as the resulting differences in the representative length and slope values, are discussed, and recommendations for proper application of the various alternatives are proposed. The alternative data reduction models are compatible with the assumptions of the D-8 method since all calculations are performed on DEM properties derived by that method.

SUBCATCHMENT LENGTH ALTERNATIVES

A subcatchment is a hillslope area that drains overland flow into the first adjacent downslope channel. Subcatchment length is often needed to estimate runoff travel distance or flow-routing distance. Two models for representative subcatchment length are proposed: the average travel distance and the average flow-path length. The average travel distance is the average of the distance from every point in the subcatchment to the first downstream channel. This average travel distance can be interpreted as the distance from the runoff-centroid of the subcatchment to the first adjacent downslope channel. The runoff-centroid is analogous to the center-of-mass, except that it represents the point of equal travel distance in upstream and downstream direction instead of the point of even mass distribution. For subcatchments that are roughly rectangular, the average travel distance is about half the subcatchment length; twice the average travel distance corresponds to the length from the drainage divide at the upstream boundary of the subcatchment to the downstream channel. The second alternative, the average flow-path length, is different because not all points in the subcatchment are considered in the length calculations. A flow-path length is defined as the distance from a divide to the first adjacent downslope channel. As such, only the distance from drainage divide cells to the downstream channel are considered in the length calculations. Note that a drainage divide is not only located at the upstream boundary of the subcatchment, but also within the subcatchment as defined by local ridges in the topography. Thus, the flow-path length is generally shorter than the average distance to the drainage divide that forms the upstream boundary of the subcatchment.

Figure 5 shows the cumulative distribution of the representative sub-catchment lengths (as calculated from the DEM) and of the map-derived lengths to the drainage divide (as estimated by manual method) for watershed WG11 in the Walnut Gulch Experimental Watershed, Tombstone, Arizona (Goodrich et al., 1994). As expected, the average flow-path length is longer than the average travel distance. This difference is attributed to the fact that the average travel distance includes the distance from all parts of the sub-catchment, whereas the flow-path lengths generally begin in the upper portions of the subcatchment where the drainage divides are predominantly located. A length-to-divide was calculated by multiplying the travel distance (distance to runoff-centroid) by 2 for rectangular subcatchment shapes or 1.5 for triangular ones. The calculated length-to-divide corresponds relatively well to the length-to-divide that was manually derived from maps. Remaining differences between the average travel distance and map-derived method are primarily related to variations in subcatchment shapes that depart from the two assumed shapes, rectangular or triangular. The interesting observation is that the proportionality factor between the average flow-path length and average travel distance is not 2 as one would

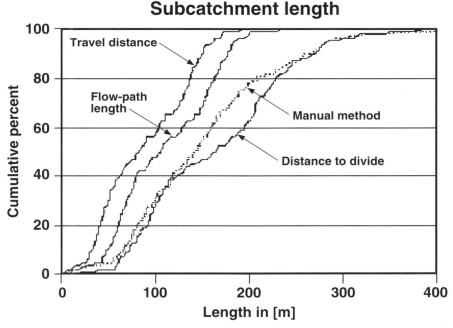

Figure 5. Frequency distribution of subcatchment length (from Garbrecht and Martz, 1999b).

have intuitively and correctly expected for rectangular subcatchments, but a value smaller than 2. This difference is related to the shape of the subcatchment and to the number and distribution of drainage divides within the subcatchment. The subcatchment shape affects the location of the runoff-centroid and the average travel distance, whereas the distribution of the drainage divides within the subcatchment affect the flow-path length. The user must select a length measure that is consistent with the analysis requirements as well as with the underlying assumptions and methods of the distributed watershed model in which the representative length is used.

SUBCATCHMENT SLOPE ALTERNATIVES

Subcatchment slope is an important variable for runoff, erosion, and energy flux calculations. Four models for representative subcatchment slope are proposed: the average terrain slope, the average flow-path slope, the average travel-distance slope, and the global slope. The average terrain slope is the average of the local slope value at every point in the subcatchment. The average flow-path slope is the average of the slope of all flow paths in the subcatchment. Again, flow paths are defined as the route followed by runoff starting at a divide and ending at the first adjacent downslope channel. The average travel-distance slope is the average of the slope from each point in the subcatchment to the next adjacent downslope channel. And the global slope is calculated as the average elevation of the subcatchment minus the average elevation of the receiving channel divided by the average travel distance.

Figure 6 shows the cumulative distribution of the representative subcatchment slopes (as calculated from the DEM) and of the map-derived slopes (as estimated by manual method) for watershed WG11. The average travel-distance slope produces the smallest slope of the four alternatives, mainly because the flatter slopes in the lower part of the subcatchments are emphasized by repeated inclusion in the travel distance of upstream points in the subcatchment. This is not necessarily a poor approximation, because it emphasizes those subcatchment areas that are subjected to larger discharges. The average flow-path slope is slightly steeper because there are fewer divides in the lower part of the subcatchment and, thus, fewer flow paths originate in the flatter portion of the subcatchment. The terrain slope produces the steepest slope because all slope values are maximum

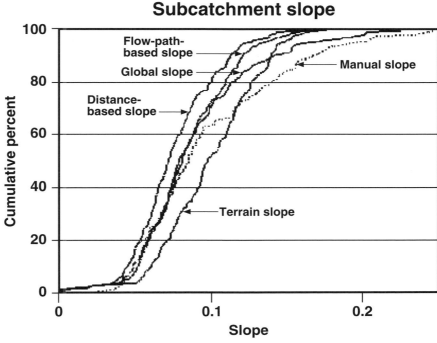

Figure 6. Frequency distribution of subcatchment slope (from Garbrecht and Martz, 1999b).

local slope values and, thus, all areas of the subcatchment, step and flat, are equally emphasized. Finally, the global slope follows the pattern of the average flow-path slope for smaller slopes and the pattern of the terrain slope for higher slopes. The better correspondence between the global and map-derived slopes is primarily owing to the fact that the map-derived slope is based on a procedure similar to that of the global slope. Thus, the better correspondence between the global and map-derived slope is not an indication of a better representative slope but only of a similarity in the underlying models. Thus, each of the four alternatives is equally valid, and users will have to select the slope value that is most appropriate for their application. In general, the terrain slope is believed to be better suited for vertical energy and mass flux calculations, whereas travel distance and flow-path lengths are more suited for runoff and erosion applications.

CONCLUSIONS   Grid-type digital elevation models are often used as a source of topographic data for distributed watershed models. This increasing popularity of DEM data is attributed in part to: (1) the cost-effective and easy access to the data; (2) a near-complete coverage of the contiguous United States at different resolutions; and, (3) advanced capabilities of geographic information systems to process the data. However, as with most data, DEMs have shortcomings and limitations that must be understood before using the data in water resources modeling applications. DEM quality and resolution are two important characteristics that can affect application results. Quality refers to the accuracy with which elevation values are reported, and resolution refers to the spacing and precision of the elevation values. DEM quality and resolution must be consistent with the scale of the application and of the processes that are modeled, the size of the land surface features that are to be resolved, the type of watershed model (physical process, empirical, lumped, and so on), and the study objectives. For many applications of physical-process-based watershed models, the USGS 30-by-30-meter DEM data have broad accuracy standards and a rather coarse resolution. The user must ensure that relevant and important topographic features are accurately resolved by the selected DEM. Custom DEMs of increased quality and resolution can be obtained at a higher cost through commercial providers. However, in practice the selection of DEM is often driven by data availability, judgment, test applications, experience, and last but not least, cost.

DEMs are often processed by GIS packages to define the configuration of the channel network, location of drainage divides, channel length and slope, and subcatchment properties. The automated derivation of such information from DEMs is faster, less subjective, and provides more reproducible measurements than traditional manual evaluation of maps. The capabilities and limitations of automated extraction of topographic data from DEMs are discussed for methodologies based on the D-8 method. Current DEM-processing packages can, based on simplifying assumptions, identify surface drainage in the presence of depressions and flat areas. However, the differentiation between real and spurious depressions, and the identification of true drainage in flat areas, remain elusive, and the user has to contend with approximations. It has also been recognized that the configuration of the drainage network, as well as its essential topologic and morphometric characteristics, are highly dependent on the definition of channel sources. The two prevailing methods for network sources in DEMs are the constant threshold area method and the slope-dependent critical support area method. The constant threshold area method appears more practical in application and does not require the estimation of a local land surface slope. Once the drainage network is identified, an automated network analysis can be performed to index channel links and junctions, establish channel connectivity, determine channel-link length and slope values, quantify the drainage network composition, and define an appropriate cascade flow routing sequence through the network.

Recent developments regarding the estimation of representative values for distributed subcatchment properties from DEMs have been presented. In particular, Data Reduction (DR) models to reduce the spatially varying length and slope values to a single representative value for the subcatchment are discussed. The representative values obtained from different DR models and map-derived methods displayed significant differences. The differences are not approximation errors, but resulted from the different definition of representative length and slope and different DR models. No one method was found to be superior to another. It is recommended that the selection of a particular representative value must be made as a function of the physical processes and circumstances in which it will be used.

Past trends and developments, the increasing quality and resolution of new DEM products, new raster-processing methodologies, as well as the expanding capabilities of GIS and linkage with traditional watershed models, lead us to believe that the use of DEMs to derive topographic and drainage data for water resources investigations will continue to increase.

REFERENCES

Band, L. E. 1986. Topographic Partitioning of Watersheds with Digital Elevation Models. *Water Resources Research* 22(1):15–24.

Cluis, D., L. W. Martz, E. Quentin, and C. Rechatin. 1996. Coupling GIS and DEM to Classify the Hortonian Pathways of Non-Point Sources to the Hydrographic Network. In *Application of Geographic Information Systems in Hydrology and Water Resources Management,* eds. K. Kovar and H. P. Nachtnebel, International Association of Hydrological Sciences Publication No. 235, 37–45.

Costa-Cabral, M. C., and S. J. Burges. 1994. Digital Elevation Model Networks (DEMON): A Model of Flow Over Hillslopes for Computation of Contributing and Dispersal Areas. *Water Resources Research* 30(6):1681–92.

Dietrich, W. E., C. J. Wilson, D. R. Montgomery, and J. McKean. 1993. Analysis of Erosion Thresholds, Channel Networks and Landscape Morphology Using a Digital Terrain Model. *Journal of Geology* 101:259–78.

Fairfield, J., and P. Leymarie. 1991. Drainage Networks from Grid Digital Elevation Models. *Water Resources Research* 30(6):1681–92.

Freeman, T. G. 1991. Calculating Catchment Area with Divergent Flow Based on a Regular Grid. *Computers and Geosciences* 17(3):413–22.

Garbrecht, J. 1988. Determination of the Execution Sequence of Channel Flow for Cascade Routing in a Drainage Network. *Hydrosoft* 1(3):129–38.

Garbrecht, J., and L. W. Martz. 1994. Grid Size Dependency of Parameters Extracted from Digital Elevation Models. *Computers and Geosciences* 20(1):85–87.

Garbrecht, J., and L. W. Martz. 1995. Advances in Automated Landscape Analysis. In *Proceedings of the First International Conference on Water Resources Engineering,* eds. W. H. Espey and P. G. Combs, American Society of Engineers, San Antonio, Texas, August 14–18, 1995, Vol. 1, 844–48.

Garbrecht, J., and L. W. Martz. 1996. Comment on "Digital Elevation Model Grid Size, Landscape Representation, and Hydrologic Simulations" by Weihua Zhang and David R. Montgomery. *Water Resources Research* 32(5):1461–62.

Garbrecht, J., and L. W. Martz. 1997a. Automated Channel Ordering and Node Indexing for Raster Channel Networks. *Computers and Geosciences* 23(9):961–66. In press.

Garbrecht, J., and L. W. Martz. 1997b. The Assignment of Drainage Direction Over Flat Surfaces in Raster Digital Elevation Models. *Journal of Hydrology* 193:204–13.

Garbrecht, J., and L. W. Martz. 1999a. TOPAZ: An Automated Digital Landscape Analysis Tool for Topographic Evaluation, Drainage Identification, Watershed Segmentation and Subcatchment Parameterization; TOPAZ Overview. U.S. Department of Agriculture, Agricultural Research Service, Grazinglands Research Laboratory, El Reno, Oklahoma, USDA, ARS Publication No. GRL 99-1, 26 pp., April 1999.

Garbrecht, J., and L. W. Martz. 1999b. Methods to Quantify Distributed Subcatchment Properties from Digital Elevation Models. Nineteenth Annual AGU Hydrology Days, Ft. Collins, Colorado. In press.

Garbrecht, J., and P. Starks. 1995. Note on the Use of USGS Level 1 7.5-Minute DEM Coverages for Landscape Drainage Analyses. *Photogrammetric Engineering and Remote Sensing* 61(5):519–22.

Garbrecht J., P. J. Starks, and L. W. Martz. 1996. New Digital Landscape Parameterization Methodologies. In *Proceedings of AWRA Annual Symposium on GIS and Water Resources,* AWRA, September 22–26, 1996, Fort Lauderdale, Florida, 357–65.

Gardner, T. W., K. C. Sasowsky, and R. L. Day. 1991. Automated Extraction of Geomorphic Properties from Digital Elevation Data. *Geomorphology Supplements* 80:57–68.

Gesch, D. B. 1994. Topographic Data Requirements for EOS Global Change Research. U.S. Geological Survey, Department of the Interior, Open-File Report 94-626, 60 pp.

Goodrich, D. C., T. J. Schmugge, T. J. Jackson, C. L. Unkrich, T. O. Keefer, R. Parry, L. B. Bach, and S. A. Amer. 1994. Runoff Simulation Sensitivity to Remotely Sensed Initial Soil Water Content, *Water Resources Research* 30(5):1393–1405.

Jenson, S. K., and J. O. Domingue. 1988. Extracting Topographic Structure from Digital Elevation Data for Geographical Information System Analysis. *Photogrammetric Engineering and Remote Sensing* 54(11):1593–1600.

Maidment, D. R., F. Olivera, A. Calver, A. Eatherral, and W. Fraczek. 1996. Unit Hydrograph Derived from a Spatially Distributed Velocity Field. *Hydrological Processes* 10(6):831–44.

Mark, D. M. 1984. Automatic Detection of Drainage Networks from Digital Elevation Models. *Cartographica* 21(2/3):168–78.

Mark, D. M. 1988. Network Models in Geomorphology. In *Modeling Geomorphological Systems,* ed. M. G. Anderson, 73–96. Chichester: John Wiley & Sons.

Martz, L. W., and De Jong. 1988. CATCH: A FORTRAN Program for Measuring Catchment Area from Digital Elevation Models. *Computers and Geosciences* 14(5):627–40.

Martz, L. W., and J. Garbrecht. 1992. Numerical Definition of Drainage Network and Subcatchment Areas from Digital Elevation Models. *Computers and Geosciences* 18(6):747–61.

Martz, L. W., and J. Garbrecht. 1995. Automated Recognition of Valley Lines and Drainage Networks from Grid Digital Elevation Models: A Review and a New Method. Comment. *Journal of Hydrology* 167(1):393–96.

Martz, L. W., and J. Garbrecht. 1998. The Treatment of Flat Areas and Closed Depressions in Automated Drainage Analysis of Raster Digital Elevation Models. *Hydrological Processes* 12:843–55.

Martz, L. W., and J. Garbrecht. 1999. An Outlet Breaching Algorithm for the Treatment of Closed Depressions in a Raster DEM. *Computers and Geosciences.* In press.

Montgomery D. R., and W. E. Dietrich. 1988. Where Do Channels Begin? *Nature* 36(6196):232–34.

Montgomery, D. R., and W. E. Dietrich. 1989. Source Areas, Drainage Density, and Channel Initiation. *Water Resources Research* 25(8):1907–18.

Montgomery, D. R., and W. E. Dietrich. 1992. Channel Initiation and the Problem of Landscape Scale. *Science* 225:826–30.

Montgomery, D. R., and E. Foufoula-Georgiou. 1993. *Water Resources Research* 29(12):3925–34.

Moore, I. D., R. B. Grayson, and A. R. Ladson. 1991. Digital Terrain Modelling: A Review of Hydrological, Geomorphological and Biological Applications. *Hydrological Processes* 5(1):3–30.

Mooris, D. G., and R. G. Heerdegen. 1988. Automatically Derived Catchment Boundary and Channel Networks and their Hydrological Applications. *Geomorphology* 1(2):131–141.

O'Callaghan, J. F., and D. M. Mark. 1984. The Extraction of Drainage Networks from Digital Elevation Data. *Computer Vision, Graphics, and Image Processing* 28:323–44.

Ostman, A. 1987. Accuracy Estimation of Digital Elevation Data Banks. *Photogrammetric Engineering and Remote Sensing* 53(4):425–30.

Quinn, P., K. Beven, P. Chevallier, and O. Planchon. 1991. The Prediction of Hillslope Flow Paths for Distributed Hydrological Modelling Using Digital Terrain Models. *Hydrological Processes* 5(1):59–79.

Ritchie, J. C. 1995. Airborne Laser Altimeter Measurements of Landscape Topography. *Remote Sensing of Environment* 53(2):85–90.

Seybert, Thomas A. 1996. Effective Partitioning of Spatial Data for Use in a Distributed Runoff Model. Doctor of Philosophy Dissertation, Department of Civil and Environmental Engineering, The Pennsylvania State University, August 1996.

Tarboton, D. G., R. L. Bras, and I. Rodrigues-Iturbe. 1991. On the Extraction of Channel Networks from Digital Elevation Data. *Water Resources Research* 5(1):81–100.

Topographic Science Working Group. 1988. Topographic Science Working Group Report to the Land Process Branch, Earth Science and Application Division, NASA Headquarters, Lunar and Planetary Institute, Houston, Texas, 64 pp.

Tribe, A. 1992. Automated Recognition of Valley Heads from Digital Elevation Models. *Earth Surface Processes and Landforms* 16(1):33–49.

U.S. Geological Survey, Department of the Interior. 1990. Digital Elevation Models: Data Users Guide. National Mapping Program, Technical Instructions, Data Users Guide 5, Reston, Virginia, 38 pp.

Wolock, D. M., and C. V. Price. 1994. Effects of Digital Elevation Map Scale and Data Resolution on a Topographically Based Watershed Model. *Water Resources Research* 30(11):3041–52.

Wolock, D. M., and J. G. McCabe. 1995. Comparison of Single and Multiple Flow Direction Algorithms for Computing Topographic Parameters in TOPMODEL. *Water Resources Research* 31(5):1315–24.

Zebker, H. A., and R. M. Goldstein. 1986. Topographic Mapping from Interferometric SAR Observations. *Journal of Geophysical Research* 91:4993–99.

Zhang, W., and D. R. Montgomery. 1994. Digital Elevation Model Grid Size, Landscape Representation, and Hydrologic Simulations. *Water Resources Research* 30(4):1019–28.

ABOUT THE AUTHORS

Jurgen Garbrecht, Ph.D. is a research hydraulic engineer at the U.S. Department of Agriculture, Agricultural Research Service. Prior to joining the Department of Agriculture, Dr. Garbrecht taught civil engineering at Colorado State University, Ft. Collins, Colorado. His areas of expertise include the identification of long-term climate variations and water resources interactions, and the use of seasonal climate forecast information in agricultural management and production; the development of digital elevation model methodologies and raster algorithms for automated characterization of topographic drainage features as needed in water resources modeling applications; hydrologic watershed system dynamics in response to global climatic and environmental change impacts, including the long-term assessment of mass and energy fluxes in large agricultural and mixed land-use watersheds; and the dynamics and interdependence of surface runoff and sediment production, transport, yield, and redistribution in large agricultural watersheds and complex drainage networks. He holds a doctoral degree from Colorado State.

Lawrence W. Martz, Ph.D., P.Geo has taught since 1984 in the Department of Geography at the University of Saskatchewan. Dr. Martz previously taught in the Water Sciences Program at the Kelsey Institute of Applied Arts and Science, Saskatoon, Saskatchewan, and earlier worked as geographer and then air photo interpreter and land analyst for the province of Alberta. He specializes in the following areas: soil erosion processes and the application of Cs-137 for monitoring sediment movement in agricultural landscapes; automated methods for the segmentation and parameterization of landscape surface hydrologic features from digital elevation models; integration of digital terrain analysis and hydrologic simulation modeling; multiscale analysis of landscape surface hydrologic features; and digital tools for the visualization of geographic information. He holds a doctor of philosophy degree in geography (University of Saskatchewan, 1987) and a master of science in geography (University of Alberta, 1979).

CONTACT THE AUTHORS

Dr. Jurgen Garbrecht
USDA, Agricultural Research Service
Grazinglands Research Laboratory
7207 West Cheyenne Street
El Reno, OK 73036
Telephone: 405-262-5291
Fax: 405-262-0133
garbrech@grl.ars.usda.gov

Dr. Lawrence W. Martz
Professor of Geography
University of Saskatchewan
9 Campus Drive
Saskatoon, SK S7N 5A5
Canada
Telephone: 306-966-5667
Fax: 306-966-5680
www.usask.ca/~martzl

Paper 2

Preparation of DEMs for Use in Environmental Modeling Analysis

WILLIAM SAUNDERS

TEXAS NATURAL RESOURCE CONSERVATION COMMISSION

AUSTIN, TEXAS

ABSTRACT

DIGITAL ELEVATION MODELS are being employed with increased frequency in the pursuit of watershed-based approaches for water resources studies and assessments. Accuracy in the digital replication of receiving stream networks and watershed boundaries is of utmost importance, particularly in cases where mathematical models may be linked to the DEM-derived output. Depending on the map scale of the DEM utilized, certain hydrologic features of a watershed may become obscured or oversimplified during the digital delineation process. The integration of a vector hydrography layer into the DEM prior to watershed delineation is one method that can correct for this loss of detail. This method is commonly referred to as "stream burning," and several such techniques are presented and discussed.

INTRODUCTION

The use of digital elevation models (DEMs) for the purposes of automated watershed and stream delineation has increased dramatically in recent years. With the advent of software algorithms for quick and efficient processing, automated approaches are quickly replacing traditional manual methods of watershed delineation. Advantages of using these automated approaches include process reliability and reproducibility, savings of time and labor, and results which, within the context of a digital domain, can easily be linked to other meaningful data sets compiled for the region under study.

As the lead environmental agency in Texas, the Texas Natural Resource Conservation Commission (TNRCC) is tasked with oversight and implementation of various water resources programs, including the Total Maximum Daily Load (TMDL) program, the Water Availability Modeling (WAM) project, and the Source Water Assessment and Protection (SWAP) program (TNRCC, 1998). These programs have significant modeling components that rely on accurate geospatial representations of the systems under assessment. For example, each program requires the accurate identification of drainage areas for literally thousands of point locations that represent water quality segment end points, surface water quality monitoring stations, wastewater discharges, USGS flow gages, water right diversions, and surface drinking water intakes/outlets. Given the volume of work required for these projects, automated methodologies for the expedient and accurate determination of drainage areas are desirable.

The ARC GRID™ module and ArcView Spatial Analyst extension are two ESRI software utilities that provide for automated watershed delineation from DEM data, through application of the FILL, FLOW-DIRECTION, FLOWACCUMULATION, and WATERSHED algorithms. Investigations (Saunders and Maidment, 1995; Mizgalewicz and Maidment, 1996) have shown, however, that the use of raster data sets for watershed and stream delineation can produce stream networks that are inconsistent with previously accepted vector representations, such as those depicted on USGS 7.5-minute quadrangle maps, the EPA's river reach files, or the digital line graphs (DLGs) of the USGS. These inconsistencies are due to problems of map scale and the lack of adequate DEM vertical resolution in areas of low relief.

One method that can help to resolve the problem of digital stream network replication involves the integration of a vector hydrography data layer into the DEM prior to watershed delineation. This process is commonly referred to as "stream burning" and can be effective in the digital reproduction of a known and generally accepted stream network. The process is not without its drawbacks, though, as the vector hydrography layer chosen for integration must be at a similar scale as the DEM and must undergo extensive preprocessing prior to the "burn-in." Other deficiencies with the process have also been noted, such as the erroneous introduction of artificial parallel streams into the drainage network (Hellweger, 1997) and the distortion of watershed boundaries delineated from the burned DEM (Saunders and Maidment, 1996).

This paper investigates various DEM manipulation techniques that involve the integration of a vector hydrography layer. The techniques examined largely represent the experiences of the author. Advantages and disadvantages of each of the techniques are discussed. The Armand Bayou and Dickinson Bayou watersheds, which are small coastal watersheds in the vicinity of Galveston Bay (figure 1), are used for this analysis. These watersheds represent the natural drainage areas to water quality stream segments that are on the State of Texas 1999 Clean Water Act Section 303(d) list (TNRCC, 1999).

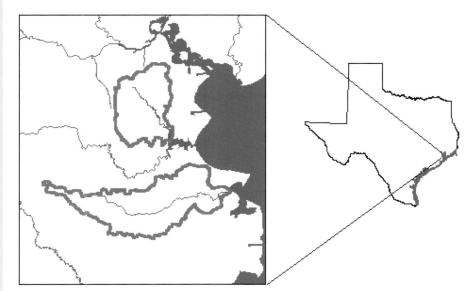

Figure 1. The Armand Bayou (north) and Dickinson Bayou watersheds in southeast Texas.

ESTABLISHMENT OF A MODEL

The concept of using DEM data to derive a representation of surface hydrography was fostered by the development of the Deterministic-8 Node (D8) algorithm (O'Callaghan and Mark, 1984). The D8 algorithm stipulates that, for a matrix (grid) representation of topography, flow moves from each node (cell) to one and only one of its eight nearest orthogonal or diagonal neighbors, and is in the direction of steepest descent. Marks et al. (1984) used this concept in creating an iterative algorithm to determine a grid of flow accumulation from a DEM. However, the number of iterations performed in this procedure was proportional to the number of grid cells in the longest flow path, and the algorithm was computationally prohibitive for large DEMs.

Jenson and Domingue (1988) introduced the concept of drainage enforcement with an automated basin delineation program that acknowledged the existence of spurious sinks in the DEM. This program allowed for the elimination of the sinks by either (a) raising the elevation of the sink point to that of its lowest neighbor or (b) lowering the elevation of the lowest saddle point surrounding the sink. By preprocessing the DEM in this manner, a fully dendritic flow accumulation network was established for all grid cells in the DEM. This concept has been implemented in the ARC GRID module and ArcView Spatial Analyst extension as part of the FILL algorithm.

The drainage enforcement concept, as discussed by Jenson and Domingue, has since been widely applied, with mixed results, for a variety of watershed and stream network delineations. Saunders and Maidment (1995) used a 3-arc-second (1:250,000-scale) DEM for a delineation of the San Antonio–Nueces coastal basin in south Texas. That analysis showed that, in the topographically diverse portion of the basin away from the coast, there was close agreement of the delineated stream network and drainage areas with the generally accepted USGS 1:100,000-scale DLG stream network and USGS Hydrologic Cataloging Units. However, in the near-coast portions of the basin, where slopes were generally flat, drainage paths were distorted and tended to short-circuit the known locations of streams and tributaries, following the DEM paths of gravitational least resistance instead (figure 2).

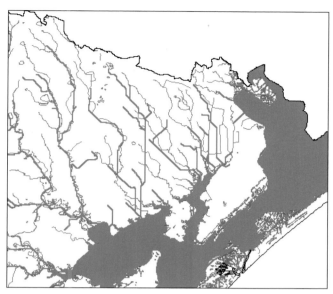

Figure 2. Digitally delineated streams (bold) in the eastern San Antonio–Nueces Coastal Basin, superimposed onto the USGS 1:100,000-scale digital line graph (Saunders and Maidment, 1995).

INTEGRATION OF VECTOR HYDROGRAPHY (STREAM BURNING)

The notion of drainage enforcement in DEMs has evolved in the years since Jenson and Domingue (1988). Taking the idea one step further, Hutchinson (1989) discussed the incorporation of stream line data into the DEM to force flow through the grid cells corresponding to the stream line network. The process would modify the elevation values of grid node points along the stream lines, constraining the elevations of interim stream line points to maintain local uniform descent. The process also ensured that the resultant delineated stream lines were located at the bottom of their accompanying valley.

To correct for the previously encountered problems in the San Antonio–Nueces coastal basin, Saunders and Maidment (1996) integrated an edited version of the USGS 1:100,000-scale DLG stream network into the 1:250,000-scale DEM prior to watershed delineation. This converted the edited stream network to a grid comprised of single-cell strings, assigned the corresponding DEM cells a fixed value of zero, and then added a fixed value (5 meters) to all off-stream cells of the DEM. Figure 3 shows the resultant improvement in digital stream delineation accuracy for the San Antonio–Nueces coastal basin.

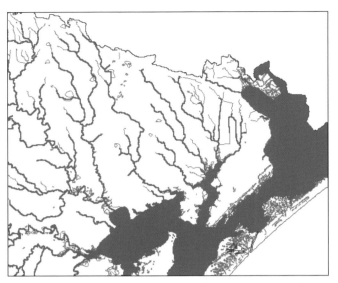

Figure 3. Digital streams created from a hydrography integration process (bold) compared with the USGS 1:100,000-scale digital line graph (Saunders and Maidment, 1996).

As part of a study on the GIS modeling of agrichemical transport, Mizgalewicz and Maidment (1996) reported on a process of DEM adjustment. The adjustment entailed integration of the 1:500,000-scale river reach file 1 (RF1) hydrography layer into a small-scale (500-meter grid cell size) DEM of the upper Mississippi–Missouri and Ohio river basins. This integration process raised all off-stream DEM cells by an arbitrary value, but retained the original DEM elevation values for the in-stream cells. The ARC GRID FILL algorithm, performed on the adjusted DEM, also provided the benefit of smoothing the resultant in-stream elevation depressions. The stream-burning process was important to this study so that flow-length estimations (and chemical travel times) would not be underestimated.

Olivera (1996) and Strager et al. (1998) also provide examples of the stream-burning technique employed by Mizgalewicz and Maidment. DiLuzio (1999) is currently employing it for all river and coastal basins in Texas through an integration of the EPA's river reach file 3 (RF3) layer into a 1-arc-second DEM.

RECENT DEVELOPMENTS

Some limitations of the above stream-burning methods have been noted. Saunders and Maidment (1996) observed distortions in watershed boundaries for drainage areas delineated with hydrography-enhanced flow direction grids (figure 4). Reed (1999) reported similar results and observed that the extent of the stream burn-in process (i.e., the difference between the elevations of stream cells and land surface cells) had a significant influence on the way that flow directions were assigned in flat areas. Hellweger (1997) noted that stream-burning processes tended to introduce parallel streams into the digital landscape in areas where the hydrography differed from the original DEM-delineated flow path by more than one grid cell. In efforts to address some of these anomalies, various DEM adjustment methodologies have been recently investigated.

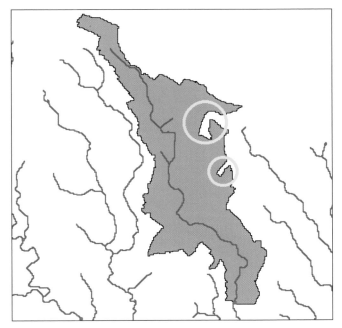

Figure 4. Watershed boundary distortions resultant from a stream-burning algorithm applied to the San Antonio–Nueces Coastal Basin (Saunders and Maidment, 1996).

Hutchinson (1996) described an adaptive interpolation process for smoothing discretization errors between adjacent DEM grid cells through an iterative resampling of the DEM. This approach has been integrated into the Australia National University DEM (ANUDEM) software package (Hutchinson, 1998), which also uses contour line data and stream line data to interpolate a smooth raster landscape surface from a data set of irregularly spaced elevation data points.

Garbrecht and Martz (1997) created the TOpographic PArameter-iZation (TOPAZ) software package to perform automated analyses of digital landscape topography. While it does not explicitly provide for integration of vector hydrography into the DEM, TOPAZ employs a method of preprocessing the DEM for watershed delineation by either filling pits and depressions or lowering the elevations of cells that obstruct flow paths. TOPAZ also eliminates flat areas from the DEM by imposing a relief based on the topography of the surrounding terrain.

In an effort to address the issue of erroneous parallel streams introduced through stream burning, Hellweger (1997) created an alternative DEM surface-reconditioning system called AGREE. AGREE notes the DEM stream cell locations that correspond to a chosen vector hydrography layer and smooths the elevations within a user-specified buffer distance of the vector layer. The extent of the elevation smoothing within the buffer is also determined by a user-specified forcing factor. After buffer smoothing is performed, the buffer zone is lowered by a fixed amount into the landscape. AGREE also provides for watershed boundary elevation reconditioning of a DEM through integration of a vector ridgeline layer. The AGREE surface-reconditioning system was applied by Quenzer and Maidment (1998) for determination of digital stream networks and drainage areas in the Corpus Christi Bay system.

More recently, Reed (1999) reported that, for hydrography integration processes, a smaller elevation differential between stream cells and land surface cells results in a more sensible assignment of flow direction values in flat areas. In accordance with that observation, a procedure has been developed that introduces a small elevation gradient in flat areas toward grid cells with a known flow direction. This process would replace the integer elevation values in the flat areas with floating point values gradually decreasing toward defined grid cells of the flow direction network.

STREAM-BURNING CONVENTIONS

The process of integrating vector hydrography into a raster elevation layer (stream burning) can be summarized in four general steps: (1) rasterization of a digitized (vector) stream network, (2) assignment of DEM elevation values to the grid cells of the raster stream network, (3) manipulation of the stream network raster cells to ensure that elevations descend toward the outlet points, and (4) introduction of a fixed elevation differential between the stream network raster cells and the land surface raster cells. Stream burning promotes the digital replication of known stream networks through the automated FLOWDIRECTION and FLOWACCUMULATION functions. This is important to ensure stream network coincidence with point locational data, such as stream gages, discharge locations, and sampling stations, that may have been located using global positioning system (GPS) techniques. The achievement of an accurate stream network is also important for water quality modeling purposes; for example, to ensure that pollutant kinetic decay parameters are applied for the appropriate stream length and period of time.

Vector hydrography integration processes can easily be misapplied. The process of preparing a vector hydrography layer for incorporation into a raster data set is tedious and fraught with pitfalls. However, a stream-burning methodology can produce accurate watershed results if a few basic, but significant, rules are adhered to:

Map Scale. A vector data layer should never be burned into a raster data layer of coarser resolution. Figure 5 illustrates the problems of burning streams into a DEM when the two layers are not of similar scale. In figure 5a, a representation of a fine-scale (e.g., 1:100,000 DLG) hydrography layer is superimposed over two conceptual grids, one coarse and one fine (e.g., a 3-arc-second [~90-meter] DEM and a 1-arc-second [~30-meter] DEM). The rasterization of the hydrography layer, for each case, is shown in figure 5b. Note that any grid cell through which a piece of the hydrography layer passes is converted to a raster stream cell. Figure 5c shows the results of a stream-thinning process, which is performed to reduce the flow network to strings of single cells. By superimposing the original hydrography over the stream cells in figure 5c, it can be seen that integration of the fine-scale vector layer into the coarse-scale DEM results in an oversimplification of the stream network. Note that one original tributary is completely omitted from the coarse network and that any point locational data that may be georeferenced to that tributary would not fall on a cell of the coarse network.

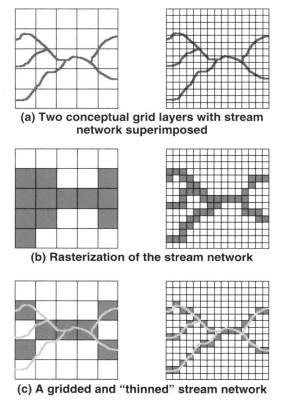

(a) Two conceptual grid layers with stream network superimposed

(b) Rasterization of the stream network

(c) A gridded and "thinned" stream network

Figure 5. Burning streams into DEMs of different scales.

Vector Hydrography Preprocessing. Prior to integration into a raster data layer, a vector stream network (e.g., EPA RF3 or USGS DLG) must be edited to create a layer that is fully dendritic (i.e., a layer that represents the drainage network with a fully connected set of single lines). In order to achieve this fully dendritic layer, the following steps must be taken:

1 All off-stream lakes (enclosed polygons) must be removed. The arcs of a hydrography layer that represent lakes actually represent the shorelines of the lakes. If these features were to be burned into the raster landscape, the result would be a number of circular trenches that would serve as local sinks. Execution of the FILL command will resolve the problem, but at the expense of additional processing time for the additional sinks.

2 In-stream lakes must be replaced with arcs that would otherwise bisect the lake. For large lakes with multiple arms or branches, a replacement dendritic transport path must be identified or created. These transport paths have been created for the National Hydrography Dataset (NHD) as "centerlines" (Dewald and Roth, 1998).

3 All coastlines must be removed. The only arcs remaining in the hydrography layer should represent drainage paths to the eventual outlets of the watershed.

4 All streams in the hydrography layer that are large enough to be represented by a left and right bank must be replaced by a single arc in the center of the stream. This is consistent with the concept of ensuring that all drainage paths are represented by a single line.

5 No bifurcations (braided streams) are allowed. For those streams where braiding naturally occurs and is depicted in the hydrography layer, a primary drainage path must be chosen and all other paths must be separated from the network. This does not mean that "secondary" drainage paths should be deleted from the layer. Alternatively (and preferably), secondary paths can be disconnected at the node where bifurcation occurs. (Note: When bifurcations are burned into a DEM, a primary drainage path is defined by the FLOWDIRECTION algorithm, which uses the D8 model. These D8-determined drainage paths can vary from the known primary paths. Allowing for the DEM elevations to determine primary flow paths defeats the purpose of stream burning.)

6 The main stems of the drainage paths must extend to the edge of the corresponding DEM. In some cases (e.g., coastal watersheds draining to a bay), this may entail extending the drainage path from the mouth of a river out into the open water. For inland basins, this will probably not be an issue.

7 No stray arcs (i.e., linear features that are not connected to the stream networks) should remain in the final layer. "Disappearing" streams that are represented in the original hydrography layer should either be eliminated or connected to the dendritic network.

8 The closest hydrographic features of adjacent watersheds (i.e., upper tributary reaches) that overlay the DEM should also be burned into the landscape. This will ultimately provide stream-burning balance near the delineated watershed boundary lines and will help to mitigate watershed boundary distortion. Preprocessing steps 1 through 7 also apply to the hydrographic features of the adjacent watersheds.

Figure 6 shows the results of applying the above preprocessing steps to the Armand Bayou hydrography layer. For comparative purposes, the edited layer is superimposed over the original USGS 1:100,000-scale DLG.

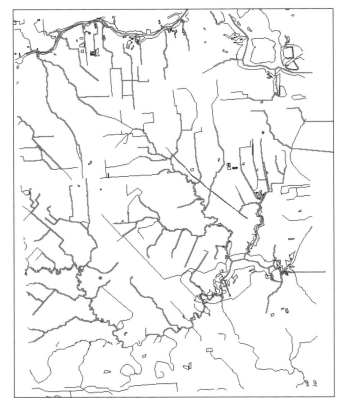

Figure 6. An edited USGS 1:100,000-scale DLG layer (red, bold) prepared for integration into a raster DEM and superimposed over the original DLG layer (dark green).

HYDROGRAPHY INTEGRATION METHODS EMPLOYED

For this analysis, four different hydrography integration algorithms were applied for stream network and drainage area delineation of the Armand and Dickinson watersheds. A baseline delineation without hydrography integration was also performed. Each algorithm was programmed in ARC Macro Language (AML™) software and executed with the ARC/INFO® version 7.2.1 ARC GRID module on a Sun® SPARC™ 20 UNIX® workstation.

Three of the four hydrography integration algorithms can be referred to as "stream-burning" algorithms. The fourth, Agree.aml, is more accurately described as a surface-reconditioning algorithm. The three stream-burning programs have a common structure: they all convert the preprocessed hydrography layer to a grid representation of the stream network, they all perform a thinning process on the grid to reduce the flow paths to the width of a single cell, they all implement a method of burning the stream grid cells into the landscape, they all execute a FILL command to ensure that no in-stream elevation depressions occur, and they all execute the FLOWDIRECTION and FLOWACCUMULATION commands. What follows is a brief description of the hydrography integration technique employed in each of the AML programs:

Fillburn.aml. This stream-burning algorithm assigns all stream grid cells the elevation values from the original DEM while assigning all off-stream cells the DEM values plus an offset value of 1,000 meters. After the FILL command is executed, the stream-burned grid is algebraically subtracted from the filled grid and the largest value in the resulting difference grid is recorded. Once that value is retrieved, the 1,000-meter offset is subtracted from the off-stream values, while the stream grid cells are reduced by the maximum value from the difference grid, incremented by one. This process minimizes the stream-burning offset required by making it only one unit deeper than the amount of the maximum fill height.

Expocurv.aml. The Expocurv.aml algorithm replaces the stream grid cell elevation values in each individual stream reach with values fitted to an exponential curve between the highest occurring elevation in the reach and the lowest occurring elevation in the reach. Additional processing is also performed to account for short reaches with no elevation change and to ensure that the highest occurring elevation values in the reach occur at the most upstream cells of the reach and the lowest at the most downstream cells of the reach.

The Expocurv.aml procedure still requires a stream-burning offset value, however. That offset is determined by first comparing the exponentially fitted elevation values with the lowest elevation values in a 3-by-3-cell neighborhood of the stream grid cells in the original

DEM. The offset value is then calculated as the maximum difference from that comparison, incremented by one. Finally, all resultant stream grid cells that have elevations less than −1 are set to −1.

Agree.aml (Hellweger, 1997). The Agree.aml algorithm offers a less computationally complicated and more processing-efficient alternative to the other stream-burning methods applied. In addition to integrating a vector hydrography layer into the landscape, Agree.aml also smooths the elevations inside a user-specified buffer zone by providing a linear fit to the grid cells from the edge of the buffer zone to the cell locations of the raster streams. Agree.aml accepts two user-specified elevation offset values, a smooth value (for integration of the vector stream network into the DEM) and a sharp value (for integration of the smoothed buffer zone into the DEM). Utilization of the sharp offset value has the effect of burning a wider trench into the digital landscape. For application to the Armand and Dickinson watersheds, the hydrography layer is buffered by 150 meters, with a smooth offset value of −5 feet and a sharp offset value of −2 feet.

Tribburn.aml. The final stream-burning algorithm represents an effort to digitally replicate the vector hydrography network while concurrently mitigating watershed boundary distortion errors. Observations of previous watershed boundary distortions show that they result from variations in the flow direction network occurring in close proximity to the grid cells burned into the digital landscape. Furthermore, the extent of the observed flow direction variations and of the resultant watershed boundary distortions are proportional to the elevation differential introduced through the stream-burning process (Reed, 1999).

With Tribburn.aml, the application of a universal offset value to all cells in the raster stream network is abandoned in favor of an iterative approach that smooths all downstream occurrences of in-stream elevations that are higher than the elevation value of the most upstream point in each reach. This process is implemented through three nested loops that consider (1) each elevation value occurring (2) in each reach (3) of each unique drainage network in the hydrography layer. Once the in-stream smoothing process is completed, the raster stream network is offset from the land surface cells by two elevation units. This represents the minimum stream-burning offset value that will still allow for digital replication of the vector hydrography network.

This Tribburn.aml procedure is processing-intensive and should not be used for delineation of stream networks and watershed boundaries when a result is desired in a short period. However, for water resources projects (e.g., TMDL analyses) that may have extended time lines, the Tribburn.aml procedure provides an alternative approach to achieving an accurate representation of the watershed drainage area.

APPLICATION TO GALVESTON BAY AREA WATERSHEDS

The Armand Bayou and Dickinson Bayou watersheds, situated just west of Galveston Bay in southeast Texas, span approximately 60 square miles and 100 square miles, respectively. For the raster delineation of these relatively small-area watersheds, 1-arc-second (~30-meter) DEMs for all 7.5-minute quadrangles encompassing the watersheds were merged together, and the resultant merged DEM was reprojected from UTM zone 15 into an Albers equal area projection. An initial FILL command was also performed to remove any spurious pits that may have been introduced as a result of the reprojection.

Prior to implementation of any hydrography integration techniques, baseline delineations of the Armand Bayou and Dickinson Bayou streams and drainage areas were performed. The resultant delineated stream networks and watershed boundaries were then compared with USGS 1:100,000-scale vector hydrography data to assess the severity of any inconsistencies. Figure 7 shows the results of this comparison for the Armand Bayou. While discrepancies between the two stream networks range from obvious (figure 7a) to subtle, there are also three watershed boundary locations that intersect the vector hydrography (figure 7b). Similar results (seven watershed boundary intersections) were observed for the Dickinson Bayou delineation. These figures illustrate the need for hydrography integration (i.e., to achieve accurate digital stream representations and watershed boundaries that are hydrologically consistent with the vector hydrography). Each of the hydrography integration techniques are assessed based on their respective abilities to achieve these goals without introducing irregularities (e.g., parallel streams and watershed boundary distortions) into the landscape.

All of the stream-burning methodologies employed were able to replicate the hydrography network accurately (although one problem with representation of a complete constructed channel was observed with the Agree.aml algorithm). Results of the various techniques are shown in figures 8 and 9. The delineated watershed boundaries for each case are depicted in bold. The red-shaded areas in the figures represent differences from the baseline delineation that were unexpected (i.e., not in the immediate vicinity of a baseline watershed boundary intersection).

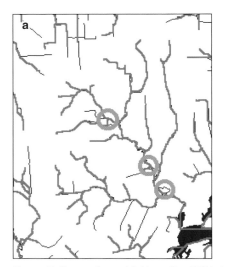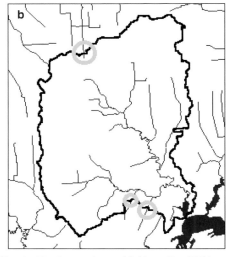

Figure 7. Comparison of (a) baseline DEM-delineated hydrography and (b) baseline DEM-delineated watershed boundary with an edited USGS 1:100,000-scale DLG layer.

The erroneous introduction of parallel streams was not a significant problem with any of the stream-burning techniques. Only one instance of parallel streams was noted for an upper watershed tributary in Armand Bayou. There were two such occurrences in Dickinson Bayou (also for upper watershed tributaries). The use of the fine-scale 1-arc-second DEMs mitigates the number of these occurrences. The lack of any significant parallel streams may also be a testament to the use of similar-scale vector and raster data sets for the analysis.

The Fillburn.aml algorithm introduced significant boundary distortions in the delineated watersheds. These are respectively shown in figures 8a and 9a for Armand and Dickinson bayous. The watershed boundary errors are due to the application of the universal stream-burning offset value to all in-stream grid cells. In the case of Armand Bayou, that value was 31 feet, resulting in an elevation differential of 32 feet between the land surface grid cells of the DEM and the in-stream cells. This result is consistent with earlier observations regarding the proportional relationship between the introduced elevation differential and the extent of flow direction variations.

Figure 8. Armand Bayou stream network and watershed boundary delineations produced using the (a) Fillburn.aml, (b) Expocurv.aml, (c) Agree.aml, and (d) Tribburn.aml hydrography integration techniques. Delineated boundaries are bold. Observed watershed distortion errors are marked in red.

There was slight improvement in Armand Bayou watershed boundary accuracy (figure 8b) with the use of the Expocurv.aml algorithm, but the Dickinson Bayou delineation (figure 9b) showed a slight loss in total drainage area with the use of that procedure. Once again, a universal stream-burning offset value was applied to all in-stream grid cells, but the method of offset calculation, comparison of the exponentially fitted in-stream elevation values with the lowest neighborhood DEM elevation values, resulted in a smaller offset for

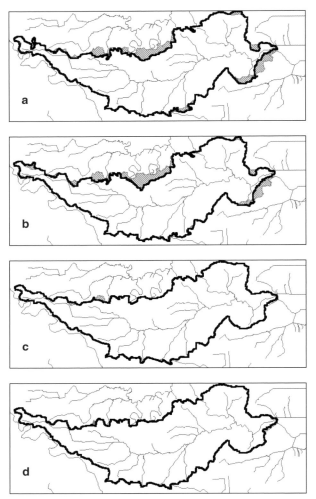

Figure 9. Dickinson Bayou stream network and watershed boundary delineations produced using the (a) Fillburn.aml, (b) Expocurv.aml, (c) Agree.aml, and (d) Tribburn.aml hydrography integration techniques. Delineated boundaries are bold. Observed watershed distortion errors are marked in red.

Armand Bayou and a greater offset for Dickinson Bayou. An additional advantage of using the Expocurv.aml algorithm was a reduction in processing time. This was the result of fewer sinks encountered during the FILL process.

The Agree.aml algorithm was the most processing-efficient of all integration techniques applied and generally produced much better conformity with the expected watershed boundary. Only three significant distortion errors were noted for Armand Bayou (figure 8c) and only two upper watershed discrepancies were observed for Dickinson Bayou (figure 9c). One of the Armand Bayou errors occurred in the eastern edge of the watershed where a tributary is close to the watershed boundary. The Agree.aml methodology of smoothing elevations within a stream network buffer zone is the source of this error. The buffer zone value selected for these delineations (150 meters) smoothed all elevations within five grid cells of the stream network cells. As this tributary is within five cells of the Armand Bayou boundary, the boundary cells in that area were smoothed and a small patch of adjacent catchment cells was included in the drainage area.

A fourth Armand boundary distortion error, observed with the Agree.aml algorithm, occurred in the southern area of the watershed and is the result of the unsuccessful integration of a constructed channel tributary there. The postintegration FILL process performed appears to have filled the in-stream cells back to the level of the land surface cells. This anomaly would have been avoided through the use of greater smooth and sharp offset values, although larger offset values would have introduced a greater potential for watershed boundary distortion.

The most accurate watershed boundary results were obtained with the Tribburn.aml algorithm (figures 8d and 9d). However, the processing cost of achieving these results was significant. While the other three techniques had processing times between 13 and 21 minutes, Tribburn.aml took six and a half hours to complete for Dickinson Bayou and more than seven hours to run for Armand Bayou. Variations in processing time depend on the density of the hydrography layer being integrated into the landscape. It should also be noted that, while watershed boundary results were notably improved over the other three techniques, some small watershed boundary distortions did occur with Tribburn.aml. Table 1 shows compiled statistics for each of the hydrography integration techniques applied.

TABLE 1. RESULTS OF VARIOUS HYDROGRAPHY INTEGRATION TECHNIQUES PERFORMED FOR STREAM AND WATERSHED DELINEATION OF ARMAND BAYOU AND DICKINSON BAYOU

Armand Bayou

Method	Stream errors	Watershed intersects	Watershed distortions	Parallel streams	Processing time
No burn	***	4	N/A	N/A	N/A
Fillburn	0	0	7	1 (upper trib)	19:22
Expocurv	0	0	6	1 (upper trib)	16:03
Agree	0	1	3	0	12:58
Tribburn	0	0	1	1 (upper trib)	7:42:52
***Significant					

Dickinson Bayou

Method	Stream errors	Watershed intersects	Watershed distortions	Parallel streams	Processing time
No burn	***	7	N/A	N/A	N/A
Fillburn	0	0	17	2 (upper trib)	21:13
Expocurv	0	0	16	2 (upper trib)	17:46
Agree	0	0	2	0	13:54
Tribburn	0	0	1	2 (upper trib)	6:28:58
***Significant					

CONCLUSIONS

Four methods of integrating stream vector hydrography into a raster data set (DEM) have been presented and reviewed. Each of the methods provides for accurate digital replication of a receiving stream network. However, subsequent watershed delineations performed with the hydrography-enhanced DEM have been shown to produce results ranging from poor to excellent. The Fillburn.aml and Expocurv.aml algorithms, which apply a universal elevation offset factor to all in-stream grid cells, produced inaccurate watershed boundaries. Application of the Agree.aml algorithm resulted in general conformity with expected results, although some watershed boundary distortions were observed. The magnitude of watershed boundary errors can be controlled by the Agree.aml user, through discriminate selection of hydrography buffer zone widths and smooth/sharp elevation offset factors. The Agree.aml algorithm provides the additional benefit of eliminating parallel streams from the hydrography-enhanced raster surface, although the use of fine-scale 1-arc-second DEM data was also shown to mitigate the number of these occurrences.

Highly accurate watershed boundary delineations were produced by the Tribburn.aml algorithm. The procedure's iterative method of adjusting in-stream elevation values on a reach-by-reach basis throughout the hydrography network effectively generated expected watershed boundary locations while minimizing the elevation offset value applied to all in-stream grid cells. The Tribburn.aml algorithm is the least processing-efficient of all the methodologies examined, however. Watershed delineations for the small Armand Bayou and Dickinson Bayou watersheds were performed 20 to 30 times slower than with the other three methodologies.

The Agree.aml and Tribburn.aml algorithms provide acceptable options for water resources modeling analyses that require geospatial data inputs to watershed and stream network models. For batch-type analyses where multiple watersheds may need to be assessed from a number of point locations in a dynamic fashion (such as an assessment of multiple water rights in a large basin), the processing-efficient Agree.aml is the more appropriate option. For cases where a single watershed delineation may be used for multiple assessments within the watershed (such as determination of statistical precipitation runoff distributions or nonpoint pollution loading assessments), the Tribburn.aml algorithm provides the most accurate delineation technique. Further investigation into the development of a process that combines the accuracy and processing efficiency of these two algorithms would be warranted.

The vector hydrography preprocessing required for use with either of these algorithms represents an intensive effort. The development of an automated technique for this processing would be an effort worthy of pursuit.

ACKNOWLEDGMENTS I would like to thank the following for continued consultation, review, and information exchange: Dr. David Maidment, Dr. Francisco Olivera, and the GIS hydrology research group at the University of Texas Center for Research in Water Resources; Dr. R. Srinivasan and Mauro DiLuzio at the Blacklands Research Center, Temple, Texas; Dr. Seann Reed at the National Weather Service Hydrologic Research Laboratory; Dr. Pawel Mizgalewicz at 3DIterralogic; Ferdinand Hellweger at HydroQual, Inc.; Jennifer Benaman at Quantitative Environmental Analysis, Inc.; and James Edmonds at the Texas Natural Resource Conservation Commission.

REFERENCES

Dewald, T. G., and K. S. Roth. 1998. The National Hydrography Dataset: Integrating the USEPA Reach File and USGS DLG. In *1998 ESRI International User Conference Proceedings,* Environmental Systems Research Institute, Inc., Redlands, California.

DiLuzio, M. 1999. Filling the Gaps Between USGS DEMs and the USEPA Reach File. In *1999 ESRI International User Conference Proceedings,* Environmental Systems Research Institute, Inc., Redlands, California. Forthcoming.

Garbrecht, J., and L. W. Martz. 1997. TOPAZ: An Automated Digital Landscape Analysis Tool for Topographic Evaluation, Drainage Identification, Watershed Segmentation and Subcatchment Parameterization; TOPAZ Overview. U.S. Department of Agriculture, Agricultural Research Service, Grazinglands Research Laboratory, El Reno, Oklahoma, ARS Publication No. GRL 97-3.

Hellweger, F. L. 1997. AGREE - DEM Surface Reconditioning System. http://www.ce.utexas.edu/prof/maidment/gishydro/ferdi/research/agree/agree.html as of July 1999.

Hutchinson, M. F. 1989. A New Procedure for Gridding Elevation and Stream Line Data with Automatic Removal of Spurious Pits. *Journal of Hydrology* 106:211–32.

Hutchinson, M. F. 1996. A Locally Adaptive Approach to the Interpolation of Digital Elevation Models. In *Proceedings: Third International Conference/Workshop on Integrating GIS and Environmental Modeling,* National Center for Geographic Information and Analysis, University of California, Santa Barbara.

Hutchinson, M. F. 1998. ANUDEM Version 4.6.2. http://cres.anu.edu.au/software/anudem.html as of July 1999.

Jenson, S. K., and J. O. Domingue. 1988. Extracting Topographic Structure from Digital Elevation Data for Geographic Information System Analysis. *Photogrammetric Engineering and Remote Sensing* 54(11):1593–1600.

Marks, D., J. Dozier, and J. Frew. 1984. Automated Basin Delineation from Digital Elevation Data. *Geo-Processing* 2:299–311.

Mizgalewicz, P. J., and D. R. Maidment. 1996. Modeling Agrichemical Transport in Midwest Rivers Using Geographic Information Systems. Center for Research in Water Resources Online Report 96-6, University of Texas, Austin, Texas.

O'Callaghan, J. F., and D. M. Mark. 1984. The Extraction of Drainage Networks from Digital Elevation Data. *Computer Vision, Graphics and Image Processing* 28:323–344.

Olivera, F. 1996. Spatial Hydrology of the Urubamba River System in Peru Using Geographic Information Systems (GIS). Center for Research in Water Resources, University of Texas, Austin, Texas.

Quenzer, A. M., and D. R. Maidment. 1998. A GIS Assessment of the Total Loads and Water Quality in the Corpus Christi Bay System. Center for Research in Water Resources Online Report 98-1, University of Texas, Austin, Texas.

Reed, S. 1999. Personal communication, National Weather Service Hydrologic Research Laboratory.

Saunders, W. K., and D. R. Maidment. 1995. Grid-Based Watershed and Stream Network Delineation for the San Antonio–Nueces Coastal Basin. In *Proceedings: Texas Water '95: A Component Conference of the First International Conference of Water Resources Engineering,* August 16–17, 1995. American Society of Civil Engineers, San Antonio, Texas.

Saunders, W. K., and D. R. Maidment. 1996. A GIS Assessment of Nonpoint Source Pollution in the San Antonio–Nueces Coastal Basin. Center for Research in Water Resources Online Report 96-1, University of Texas, Austin, Texas.

Strager, M. P., J. J. Fletcher, and C. B. Yuill. 1998. Hydrologic Modeling for Acid Mine Drainage in West Virginia. In *1998 ESRI International User Conference Proceedings,* Environmental Systems Research Institute, Inc., Redlands, California.

Texas Natural Resource Conservation Commission. 1998. Technical Paper Number 3: Digital Elevation Modeling for the WAM and Other TNRCC OWRM Projects. TNRCC, Austin, Texas.

Texas Natural Resource Conservation Commission. 1999. State of Texas 1999 Clean Water Act Section 303(d) List and Schedule of Development for Total Maximum Daily Loads, SFR-58/99. TNRCC, Austin, Texas.

ABOUT THE AUTHOR

William Saunders is an engineering specialist on the Total Maximum Daily Load (TMDL) Team of the Texas Natural Resource Conservation Commission (TNRCC). For four years at the commission, Mr. Saunders has been involved with water quality modeling, waste load evaluations, application of GIS methodologies, and TMDL assessments. Before joining the TNRCC in 1996, he spent 11 years as an electronics engineer at the Naval Underwater Systems Center in Newport, Rhode Island, and two years as a research assistant at the University of Texas Center for Research in Water Resources. He holds a B.S. in engineering physics from the University of Maine at Orono (1982) and a M.S. in civil engineering from the University of Texas (1996).

CONTACT THE AUTHOR

William Saunders
Engineering Specialist
Texas Natural Resource Conservation Commission
MC-150
P. O. Box 13087
Austin, TX 78711-3087
Telephone: 512-239-4535
Fax: 512-239-1414
bsaunder@tnrcc.state.tx.us

Paper 3

Source Water Protection Project: A Comparison of Watershed Delineation Methods in ARC/INFO and ArcView GIS

ALBERT PEREZ

ENVIRONMENTAL AFFAIRS OFFICE

WASHINGTON STATE DEPARTMENT OF TRANSPORTATION

OLYMPIA, WASHINGTON

ABSTRACT

THE ENVIRONMENTAL AFFAIRS OFFICE (EAO) of the Washington State Department of Transportation (WSDOT) has been contracted by the Washington State Department of Health (WSDOH) to delineate watersheds representing Source Water Protection Areas for drinking water systems supplied by surface water sources throughout the state. In 1998, preliminary delineation of these watersheds was performed using statewide digital elevation models and the hydrologic modeling functions of the ARC GRID module of ARC/INFO software. This year, delineations are being revised using the Watershed Delineator, an ArcView GIS extension produced by ESRI and the Texas Natural Resources Conservation Commission. A comparison is made between last year's delineation methods and those used this year. Applications of EAO's watershed delineation tools are also discussed.

1996 amendments to the federal Safe Drinking Water Act require states to develop Source Water Protection Programs with pollution prevention measures. One of the required elements is the delineation of source water protection areas, which are defined as areas in which contaminants could reach intakes of water systems. For surface water systems, this area is either an entire watershed specific to an intake point, or, for very large areas, some subset wherein contaminants will not be sufficiently diluted or otherwise degraded before entering the water system intake. Although Washington State has been divided into 62 separate Watershed Resource Inventory Areas, none of these divisions is specific to a specific water intake location (figure 1).

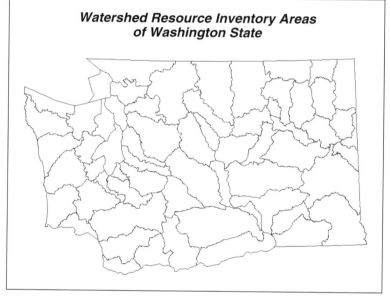

Watershed Resource Inventory Areas of Washington State

Figure 1

In 1997, the Environmental Affairs Office (EAO) of the Washington State Department of Transportation (WSDOT) was contracted by the Washington State Department of Health (WSDOH) to delineate Source Water Protection Areas for drinking water systems supplied by surface water sources. This effort came at a time when EAO was strengthening its focus on a "watershed approach" to meeting various transportation challenges, including wetland mitigation banking, fish passage barrier removal, stormwater management, flood prevention and management, hazardous materials management, and participation in local and regional transportation, watershed, and land-use planning processes. EAO is strongly interested in the development of watershed delineation tools that are accurate and timely, and which can be used by persons who do not have extensive GIS backgrounds.

The second delineation effort commenced in 1998, with the goal of using more accurate data on surface water intake locations to improve the accuracy of the delineations.

DATA SOURCES

The initial delineation effort used pourpoints representing surface drinking water system intake locations, which had been digitized from a 1:1,000,000-scale paper map of Washington State. When these points were initially digitized, locational accuracy was estimated to be "about a mile." Subsequent comparison with global positioning system (GPS) data revealed some of these points to be off by as much as 20 miles, which translates to a major difference in watershed area (figure 2).

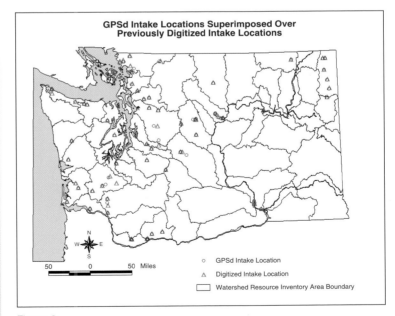

Figure 2

Elevation grid data for the entire state was obtained from the Washington State Department of Natural Resources (WADNR). This data had been derived from approximately 1,500 30-meter USGS and

Forest Service digital elevation models (DEMs), which had been filtered to remove horizontal banding and negative z-values, and then merged into a single 325-MB statewide grid covering more than 68,000 square miles (figure 3). The DEM has 12,804 rows and 19,387 columns (about 248 million cells).

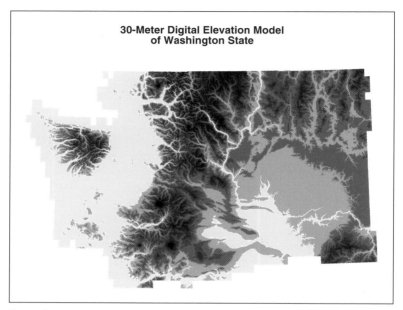

Figure 3

COMPARISON OF METHODOLOGIES

Two basic approaches were compared. The first involved the writing of ARC Macro Language (AML) scripts to utilize the hydrologic modeling functions of the ARC GRID module. The processing steps involved are described in table 1.

The second effort, performed this year, used the Watershed Delineator (written by ESRI and the Texas Natural Resource Conservation Commission) to accomplish roughly the same tasks. The Watershed Delineator is an ArcView GIS extension that requires the ArcView Spatial Analyst extension to be installed as well. This extension has a much simpler interface than the command lines in ARC GRID; all functions are accessible through buttons and menus. Each data set, starting with the raw DEM, is preprocessed by running it in order through each of the menu choices under the Hydro menu (figure 7).

The result is a group of preprocessed data sets that are ready for either interactive or batch delineation of watersheds for points, lines, or polygons of interest.

If processed grid data sets for FLOWDIRECTION and FLOWACCUMULATION already exist, they can be used directly by the Watershed Delineator.

TABLE 1. WATERSHED DELINEATION METHODOLOGY USING THE ARC GRID MODULE

1	Convert the raw DEM to ARC GRID format (figure 3).
2	Run FILL on the elevation grid to fill sinks. This will remove small imperfections in the data and enable FLOWDIRECTION to run properly. The difference between this grid and the raw DEM grid is generally not visible.
3	Run FLOWDIRECTION to create a grid of flow direction from each cell in the elevation grid to its steepest downslope neighbor.
4	Run FLOWACCUMULATION to create a grid of accumulated flow to each cell from all other cells in the FLOWDIRECTION grid.
5	Select a subset of cells with a threshold FLOWACCUMULATION value, to define a drainage network (figure 4).
6	Convert surface water intake locations into watershed outlet (pourpoint) grid cells. Ensure that pourpoints are located directly over a grid cell from the drainage network (figure 5).
7	Run WATERSHED to delineate the watershed associated with each surface water intake (figure 6).

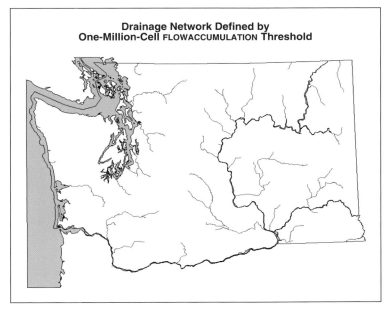

Drainage Network Defined by One-Million-Cell FLOWACCUMULATION Threshold

Figure 4

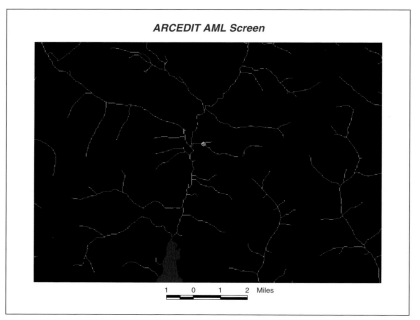

Figure 5

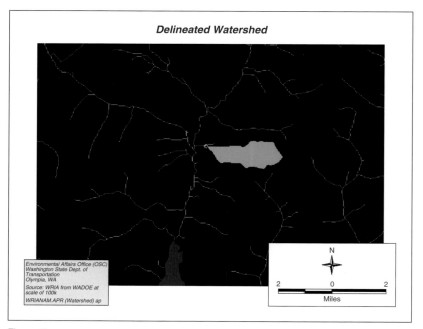

Figure 6

However, instead of using preexisting vector data sets (such as streams and watersheds), these should be generated with the Watershed Delineator, due to the specific database design required by its Avenue™ scripts.

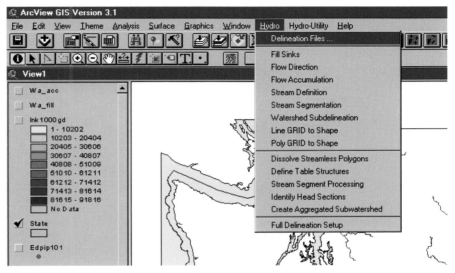

Figure 7

As a substitute for the digitized intake points used in the initial delineation effort, latitude/longitude GPS locations for the intakes were imported as an event theme into ArcView GIS, then reprojected to the same projection as the statewide elevation grid. Since intakes for many of the water systems are located along small streams or creeks, a finer-resolution, one-thousand-cell threshold stream grid was then brought in to serve as a background against which to locate the GPS points for the intakes. Although it may seem counterintuitive to get accurate GPS data and then move those points to correspond with a stream grid, it must be remembered that, for watershed modeling purposes, the DEM is the context in which we are working. When calculating a watershed, the outlet, or pourpoint, must be located exactly over a grid cell that is a member of the defined stream grid. If a pourpoint is not located directly over a stream, a small, incomplete watershed will be delineated, consisting of just a few cells.

For this reason, it is advisable to confirm the location of a pourpoint along a stream with other data about its location. In this case, tabular data from WSDOH was used to get the name of the lake, river, stream, or creek that the water system was drawing from. Another background layer, the USGS 7.5-minute quad maps in digital raster graphic format, was especially helpful in this process. For very small streams that are not shown on 7.5-minute quad maps, local site plan maps for the water system were used.

Zooming in to the level at which grid cells are visible gives the added benefit of being able to recognize parts of the DEM-derived streams that are clearly inaccurate due to insufficient topographic relief. Typically, errors are evidenced by right angles, or by long, straight stream sections, both of which are unnatural. Delineation of watersheds for intake points downstream of these types of grid-derived streams is prone to error, unless elevation data for the area in question is first replaced with elevation data of a finer resolution.

With the ARC GRID AML method, delineation of each intake's watershed required ARC GRID to process the entire statewide elevation grid. In order to speed up this year's delineation process, a stream grid was created using a one-million-cell FLOWACCUMULATION threshold. This enabled the creation of watershed areas which, although large, were smaller than the entire state. To delineate a small watershed falling within one of these larger ones, we extracted a portion of the FLOWDIRECTION grid that corresponded to the larger watershed, but which still encompassed the smaller subject watershed. Although this step could have been done with the SETWINDOW command in ARC GRID, the Watershed Delineator's DEM extraction tools greatly simplified the task. Delineation of the smaller watersheds then proceeded much more quickly using the finer-resolution (one-thousand-cell) stream grid.

Another way in which the Watershed Delineator speeds the delineation process is through the creation of aggregated subwatershed polygons, which are created as one of the menu choices under the Hydro menu. Although this requires some additional preprocessing, it allows much quicker point-specific watershed delineation.

Once the pourpoint has been located over the closest stream grid cell, the Watershed Delineator's "Delineate on a Point" tool can be used to delineate the intake-specific watershed. This method is much quicker than the standard delineation function in ARC GRID, since delineation is only performed up to the next upstream confluence. A premerged watershed polygon, created when subwatershed polygons were aggregated, represents the whole upstream area draining to that confluence. This aggregate polygon is then added to the newly calculated subwatershed to create the full watershed area for the intake point.

The real power of the Watershed Delineator is revealed when all of the pourpoints are accurately located over gridded streams. At that point, the Batch Watershed Delineation tool can be used to calculate all the watersheds for each of the intake points at the same time. This results in a substantial savings of time over the ARC GRID AML method, in which watershed grids were generated individually.

PROCESSING TIME COMPARISON

As table 2 shows, the basic preprocessing tasks of the Watershed Delineator should optimally be performed in ARC GRID on a workstation with a fast processor. When the preprocessing is completed, the actual watershed delineation is much faster with the Watershed Delineator extension. If only one Microsoft® Windows NT® workstation is available, preprocessing large grid data sets can bog down the processor, which slows down other Windows NT applications significantly. A better approach is to have a separate machine (either Windows NT or UNIX) dedicated to processing large grid data sets.

TABLE 2. COMPARISON OF PROCESSING TIMES FOR SELECTED WATERSHED DELINEATION TASKS

Workstation description	Operating system	Software	Elapsed time (hr:min:sec)			
			FLOW-DIRECTION	FLOW-ACCUMULATION	10,000-cell stream threshold	Watershed delineation
IBM® RISC System/6000® Model 90	AIX® 3.2	ARC GRID	2:55:00	20:39:00	0:06:19	2:57:00
Sun Ultra™ 60 Model 2360	Solaris™ 5.7	ARC GRID	**1:07:00**	**8:31:00**	**0:05:01**	2:49:00
Gateway™ P6 266 MHz, 384 MB RAM	Windows NT 4.0	Watershed Delineator	1:19:39	15:28:00	0:06:36	**0:00:53**
Note: Each command was compared using the same data set. Fastest times are shown in bold print.						

Table 3 shows a summary of the differences between the ARC GRID AML method of watershed delineation and the Watershed Delineator.

TABLE 3. SUMMARY COMPARISON OF DELINEATION METHODS

ArcView GIS with ArcView Spatial Analyst and Watershed Delineator extensions	ARC GRID AMLs
Although familiarity with ArcView GIS is essential, advanced GIS knowledge is not required. Easy windowing interface allows quick choice of which processes to invoke.	User must be familiar with ARC Macro Language (AML) and the use of the hydrologic modeling functions. Processing options must be written in AML code.
In addition to FILL, FLOWDIRECTION, and FLOWACCUMULATION, requires additional pre-processing to create stream segments and to aggregate subwatersheds. This allows much quicker point-specific delineation.	Preprocessing involves only FILL, FLOWDIRECTION, and FLOWACCUMULATION.
The Watershed Delineator extension includes many useful features, including: • Partial DEM extraction • Interactive and batch delineation of watersheds from points, segments, and polygons • Subwatershed aggregation	Basic functionality: defines streams and delineates watersheds.
ArcView GIS environment allows quick comparison of data sets. Important for aligning pourpoints with gridded streams. Failure to do this will cause abbreviated watersheds that are not indicative of area contributing to intake.	Background layers for comparison can be brought up with an AP AML from within ARCEDIT. Pourpoint verification step allows user to cancel processing of watersheds for pourpoints that are not aligned over a gridded stream.
Allows batch delineation. However, this can lead to errors, unless each point's location is verified first.	Requires additional AML to be written.
Documentation needs to be improved. A stepwise tutorial would be valuable.	AMLs constitute documentation in themselves, in addition to any remarks that may be included.

FUTURE APPLICATIONS

WSDOH plans to make the results of this work available to other agencies and the general public via the Internet, in order to facilitate protection of drinking water supplies throughout the state.

Watershed delineation tools will be used by WSDOT and WSDOH to help prevent, mitigate, control, and remediate transportation-related adverse impacts to surface-derived drinking water. For example, this tool can be used immediately to augment the existing environmental documentation process for proposed transportation projects. Watershed delineation will also be helpful in determining best management practices for handling stormwater runoff from highways and other WSDOT property.

Watershed boundary data may be used in conjunction with other data layers, such as endangered species habitat, wetlands, storm water outfall locations, in-stream water quality segments, and hazardous materials sites, to conduct an initial assessment of potential environmental impacts associated with Washington's 20-year State Highway System Plan. In addition, these watershed delineation tools can be used to delineate location-specific watersheds for DOT construction projects. For example, if WSDOT receives a permit from a local jurisdiction with watershed-related mitigation requirements, the watershed can be generated on demand for the specific project in question.

Recently, two events have added to EAO's interest in enhancing its capability to delineate watersheds. Referendum 49, passed by the voters of Washington State, authorized $2 billion in transportation improvements in the coming two years. This need to build is juxtaposed against the recent listing of six species of salmon under the Endangered Species Act. Knowledge of where the potential effects of highway improvement projects will occur will be critical to the timely delivery of these essential transportation services.

CONCLUSIONS

The Watershed Delineator is a good example of a user-friendly analysis environment that can be created with the hydrologic modeling tools available in ArcView Spatial Analyst, and represents a significant step forward from the ARC GRID AML approach. However, a good understanding of the quality of the input data, as well as a firm knowledge of the hydrologic modeling processes that the software is applying, are good insurance against erroneous results that can be obtained through misapplication of this useful tool.

ACKNOWLEDGMENTS

Funding for this project was provided by the Washington State Department of Health. Valuable assistance with technical issues was provided by the Washington State Department of Ecology, Washington State Department of Health, and Washington State Department of Natural Resources, and by ESRI's offices in Olympia, Washington, and Redlands, California.

ABOUT THE AUTHOR

Albert Perez is a GIS analyst with the Environmental Affairs Office of the Washington State Department of Transportation. Mr. Perez has worked in the past as a private planning and GIS consultant, and for various other state and county governments. He has a master's degree in urban and regional planning from the University of Hawaii at Manoa.

CONTACT THE AUTHOR

Albert Perez
GIS Analyst
Environmental Affairs Office
Washington State Department of Transportation
310 Maple Park East
MIS 47331
Olympia, WA 98504-7331
Telephone: 360-705-7582
Fax: 360-705-6833
pereza@wsdot.wa.gov

Paper 4

DEM Preprocessing for Efficient Watershed Delineation

DEAN DJOKIC AND ZICHUAN YE

ESRI

REDLANDS, CALIFORNIA

ABSTRACT

GIS IS WIDELY USED to support water quantity and quality studies. GIS and digital elevation models (DEMs) can be used to perform watershed delineation to a point, a reach, or an area of interest, which is usually one of the first steps in such studies. This paper presents the methodology that preprocesses the DEM in order to facilitate interactive watershed delineation. As a result, watersheds can be delineated quickly and with consistent time response, regardless of the DEM size or the size of the resulting watershed. This methodology is used as a basis for interactive basin development in GIS tools for HMS model support.

Watershed delineation is one of the most commonly performed activities in hydrologic and environmental analyses. Digital elevation models (DEMs) provide good terrain representation from which watersheds can be derived automatically using GIS technology. The techniques for automated watershed delineation have been available since the mid-eighties and have been implemented in various GIS systems and custom applications (Garbrecht and Martz, 1999).

These techniques are independent of the DEM resolution and their use has become more popular with the advent of higher resolution DEMs that allow more detailed terrain representation and analysis. These techniques are also attractive because increased desktop-computing power now allows the complex operations involved in this process to be performed locally and fast enough. These two factors together with the increased emphasis in industry for watershed-wide solutions are placing increasing demands on the automated systems to provide results quickly.

In general, the traditional approach in automated watershed delineation required high-end GIS and often resulted in long processing times (hours) and times that varied with respect to the location of the point of interest (Perez, 1999). For example, if the point of interest was close to the overall watershed boundary, the processing time would be faster than if that point was close to the watershed outlet. This was not conducive to interactive use of the methodology and limited its use to GIS shops. It also limited overall use of the methodology since the people needing the watershed boundaries were not willing to wait, sometimes for days, to get results from the GIS group (just to realize that some other points would be of interest and so have to wait a few more days for the results).

This paper presents a methodology for DEM preprocessing that provides the basis for fast (under a minute) and consistent watershed delineation on DEMs of any resolution and size using desktop GIS technology. This methodology will be referred to as the Fast Watershed Delineation method (FWD). It was initially developed in 1997 (Djokic et al., 1997) as part of the watershed delineation project developed for the Texas Natural Resource Conservation Commission.

The FWD methodology is based on the following key factors:

- The derived terrain properties (e.g., flow direction and flow accumulation) do not change unless the DEM changes, and are a function not of watershed delineation for any particular point, but rather of the DEM itself, and thus do not have to be linked to the individual watershed delineation operation (and should not be).

- The time needed to delineate a watershed is a function of the grid (DEM) size—the more cells in a grid, the longer the processing time.

- The key property of a watershed boundary is that it completely and uniquely defines the area from which the surface water drains to the watershed outlet. Any point outside of that area does not contribute to the flow at the outlet point, and thus is not of direct interest for problems related to that watershed.

The FWD methodology consists of two preprocessing steps and a technique for using the preprocessed data. They will be described in the following sections.

Step 1: Determination of derived DEM properties. In this step the key terrain derivatives (flow direction and flow accumulation grids based on the elevation grid) are computed and saved, so when needed for individual watershed delineation they do not have to be computed again, but are directly accessed. This means that these operations need to be performed only once for a given DEM (if the DEM changes, then these operations need to be repeated). This also allows the computations to be performed on a different computer than the one on which the watershed delineation will be performed. Since these computations are some of the most resource intensive, they can be delegated to the most powerful machines, or to dedicated computers. Once the processing is done, the results can be sent to other machines where the less intensive and more interactive operations will be performed.

Step 2: Initial, arbitrary delineation. In this step the initial, arbitrary watershed delineation that will serve as the basis for interactive delineation is performed. Figure 1 depicts a large watershed with its stream network. This watershed will be subdivided into several subwatersheds to increase the performance of interactive delineation. Since the initial delineation is arbitrary, any technique producing acceptable results can be applied, so in general, the simplest GIS procedures can be used for this task.

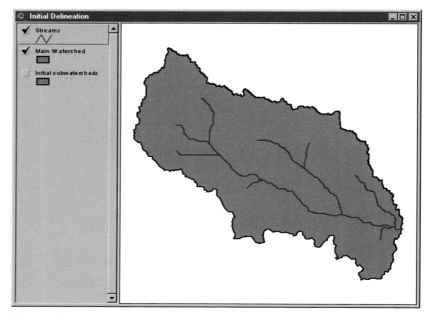

Figure 1. Main watershed with the stream network.

The most straightforward GIS technique for subwatershed delineation consists of the following steps (described in more detail in ESRI, 1997 and Olivera and Maidment, 1999):

1 Determine flow direction grid (DEM derived property).

2 Determine flow accumulation grid (DEM derived property).

3 Specify a stream threshold on the flow accumulation grid. This operation will identify all the cells in the flow accumulation grid that are greater than the provided threshold. A new grid (a stream grid) is formed from those cells. This grid will be an indication of the drainage network. The threshold value in this process does not have any particular geomorphologic meaning through which we are trying to identify the real drainage network, but instead is used as a means for watershed partitioning. Higher thresholds will result in a less dense network and fewer internal subwatersheds, while lower thresholds will result in a denser network and more internal subwatersheds. The choice of threshold value and its impact on the delineation performance will be discussed later.

4 The stream grid is converted into stream segments, where each head segment and segment between the junctions has a unique identifier.

5 Subwatersheds (in grid format) are defined for each of the stream links in the stream link grid.

6 Subwatershed and stream grids are vectorized to produce subwatershed and stream polygon and polyline themes respectively. Additional vector processing might be needed to clean up the data and ensure correct connectivity and directionality.

An overview of the process is presented in figure 2.

The shaded elements in figure 2 are the actual data used to support the interactive delineation. All other data are intermediate results and are not necessary for the watershed delineation process (although they might be needed for other hydrologic operations, e.g., the DEM itself may not be needed for the watershed delineation but is needed for elevation extraction). Figure 3 depicts the initial watershed broken down into a number of nonoverlapping and all-inclusive subwatersheds after the described procedure has been applied. In this example, there are 15 subwatersheds.

In addition to the polygon layer of the nonoverlapping subwatershed polygons, another polygon theme is created that contains, for each subwatershed, the polygon of the contributing area at its upstream end (figure 4). This so-called merged watershed polygon theme is constructed by aggregating all the upstream subwatersheds from each initial subwatershed. All of these polygons are stored in a single theme, so they overlap. The role of this polygon theme will be explained in the next section. Generation of the merged watershed theme concludes the DEM preprocessing steps.

Both the initial delineation and the upstream watershed merging are a one-time operation and can be performed independently of the actual watershed delineation in a manner similar to the determination of derived DEM properties. It is suggested that these operations (or at least the grid-processing operations) be performed at GIS shops that are equipped for high-volume data processing and have extensive experience with DEM manipulation for hydrologic purposes. The processed data can then be published for use by general users who can skip many of the preprocessing tasks and be assured of the quality of the published work. Examples of such data sets are the Oklahoma 1:100,000-scale DEM (Cederstrand and Rea, 1995) and HYDRO 1K Elevation Derivative Database (USGS et al.).

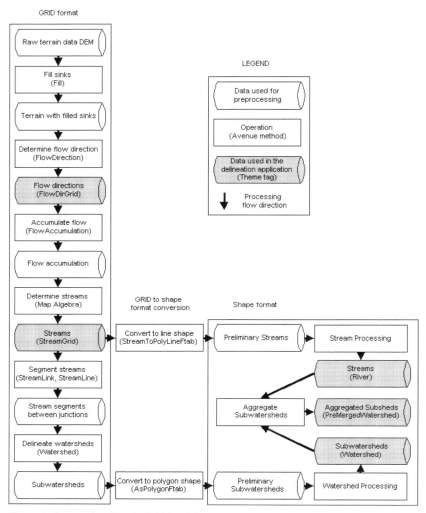

Figure 2. Processing flow for initial, arbitrary subwatershed delineation (ESRI, 1997).

The preprocessed data can be hidden from the end user if necessary. Usually, at least the stream network and the overall watershed boundary are displayed to provide the user with a visual reference, but other sources can be used for that purpose as well (such as DOQQs or scanned and georeferenced topo maps).

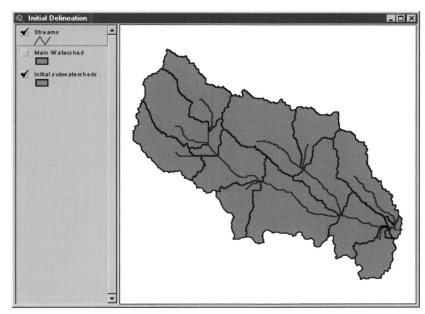

Figure 3. Subwatersheds with the stream network.

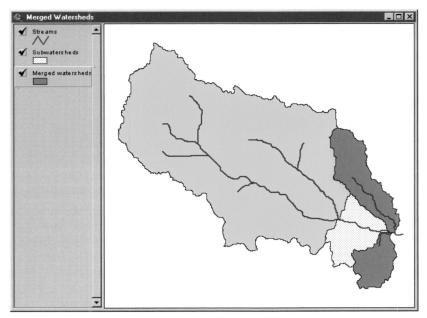

Figure 4. Example of merged upstream watershed. The dotted polygon is the subwatershed of interest. The light-colored polygon to its left is the merged polygon, while the dark polygons to its right are the areas of the overall watershed not draining into it.

TECHNIQUE FOR INDIVIDUAL WATERSHED DELINEATION

When a total drainage area to a point needs to be identified, the following procedure is performed:

1 After the user defines the point of interest, the subwatershed polygon in which this point resides is identified (figure 5).

2 The flow direction grid under the subwatershed polygon is extracted from the existing preprocessed flow direction grid.

3 The area draining into the specified point is determined from the extracted flow direction grid using an intrinsic GIS function that uses flow direction grid and outlet point grid as inputs. This will return only the grid of the area contained in the subwatershed polygon defined in step 1 (figure 6). Since the extracted grid is much smaller than the whole grid, the operation is performed significantly faster. The result of this operation is a grid that is converted into a polygon (referred to as the local drainage area polygon).

4 The polygon identifying the drainage area at the upstream end of the subwatershed identified in step 1 is retrieved from the merged watershed theme. This polygon and the local drainage area polygon are merged to form the complete drainage area for the point of interest (figure 7).

Implementation of this technique produces consistent execution times since only two polygons need to be added for any selected point (no addition is needed for points in head watersheds). There is a little variation owing to different sizes of the subwatershed in which the point might reside (a function of the threshold value used during the preprocessing and the shape characteristics of the watershed).

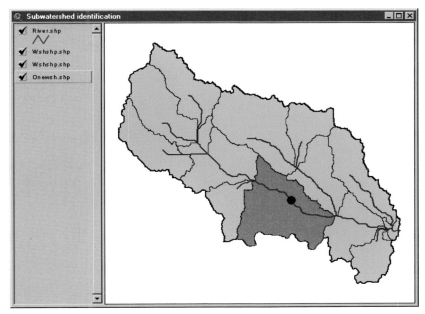

Figure 5. Identifying the subwatershed containing the point of interest.

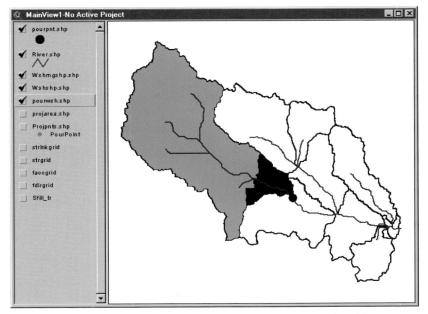

Figure 6. Identifying the local drainage area polygon (black polygon) and the upstream drainage area (lighter polygon).

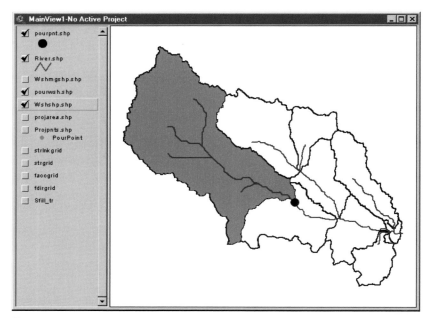

Figure 7. Final watershed for user-defined point.

TECHNIQUES FOR OTHER TYPES OF WATERSHED DELINEATION

The same principle can be used to delineate other types of watersheds. One common type of watershed delineation is determination of the surface area draining into a stream reach. This area does not include the upstream area draining into the upper end of the reach (figure 8). Determination of such a subwatershed is a simple application of point delineation. The following process is performed:

1 Delineate watershed for the upstream end of the reach.

2 Delineate watershed for the downstream end of the reach.

3 Take the difference of the two watershed polygons.

This operation takes slightly more than twice the time it takes to compute a point watershed (due to the computation of the difference between the polygons).

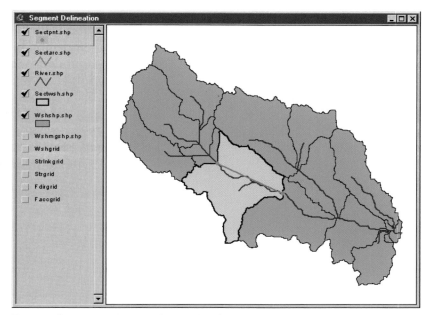

Figure 8. Segment delineation. Notice that if the segment contains a tributary stream, the contributing area of the tributary will be included in the segment watershed.

Another interesting watershed type is the area that drains through another area (as opposed to a point or a stream segment). This might be used to identify the surface area from which overland water washes over an area of interest (e.g., a city or a fire-damaged area). Figure 9 shows an example.

The technique for polygon delineation is slightly different than for the point and segment delineation, but is based on the same principle. The following steps are performed:

1 Take the original polygon of interest and split it into separate polygons residing completely within the initial subwatersheds (if necessary).

2 For each of the split polygons, identify the subwatershed it resides in and extract the flow direction grid for that subwatershed polygon.

3 Determine the area draining into the split polygon from the extracted flow direction grid using the intrinsic GIS function that uses the flow direction grid and the split polygon grid as inputs. This will return only the grid of the area contained in the subwatershed polygon.

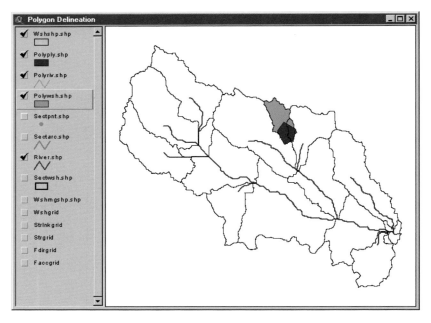

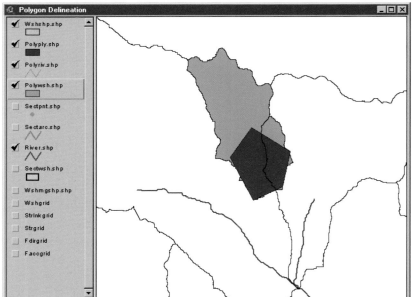

Figure 9. Overview and detail of polygon watershed delineation. Notice how the polygon of interest (dark area) spans the two initial subwatersheds. The light area indicates the watershed contributing to the flow over the polygon.

4 Add drainage area polygons for all split polygons to form the final watershed (if necessary).

Notice that if the polygon intersects the stream system, the returned watershed will not include the area that drains to the upper intersection of the polygon with the stream. Physical interpretation is that the delineated watershed contains the area that directly contributes to the flow over the polygon, not through the polygon.

PERFORMANCE ISSUES

Implementation of the FWD methodology results in a consistent watershed delineation processing time. Figure 10 presents a graph showing the relationship between the processing time for a single-point watershed and the DEM grid size without the FWD methodology. With the FWD methodology, the processing time varied between six and nine seconds regardless of DEM size. The computations were performed on the Windows NT platform using a Pentium® II 330-MHz processor and 128 MB of RAM.

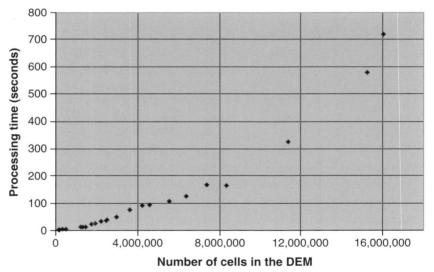

Number of cells in the DEM

Figure 10. Relationship between processing time for a point watershed delineation and the DEM size (number of cells) without the FWD methodology (Pentium II 330-MHz processor and 128 MB of RAM).

Perez (1999) reports that for a DEM of approximately 250 million cells, processing of a single watershed took almost three hours using the traditional method (no FWD) on a Sun Ultra 60 Model 2360 workstation. For the same DEM, with the FWD methodology, processing took about one minute on a Pentium 266-MHz machine. Actual performance for the traditional method will depend mostly on the size of the DEM (number of cells) and the location of the point of interest, and also on the shape of the DEM.

For the traditional method, less preprocessing is required (only the flow direction grid is needed), so if only a handful of points ever needs to be delineated, it is more feasible to use the traditional method. The preprocessing time needed to prepare the data for the watershed delineation technique can easily run into hours, and for large DEMs, even into days (Perez, 1999), but these operations are done only once for a given DEM. After preprocessing, the FWD methodology can be implemented on any computer without additional preprocessing.

AFFECT OF THRESHOLD SIZE ON THE DELINEATION PROCESS

The selection of the threshold affects watershed delineation in two ways. The first is the speed of delineation. If a large threshold is selected (relative to the size of the DEM), then few initial subwatersheds will be generated and the speed-up process will not be as efficient since the extracted grids will still be large (although the delineation will still be significantly faster than without the preprocessing).

The threshold selection and initial watershed subdelineation can and should be an iterative process. The threshold definition (stream definition) is performed after the flow direction and flow accumulation grids are determined (the two most time-consuming operations), so not all the preprocessing steps need to be performed again. Also, the vector preprocessing does not have to be performed until the final threshold value is selected. The following process can be implemented:

1 Perform flow direction and flow accumulation computations.

2 Select a threshold depending on the desired performance. Based on the initial experiences, if any of the subwatersheds does not exceed about 500,000 cells, individual watershed delineation for a point will be performed in less than 10 seconds on today's typical desktop computers (Pentium 300 with 128 MB of RAM). A good threshold starting value can be 1/500 the number of cells in the DEM.

3 Using this threshold, perform other grid-preprocessing steps until the grid of initial subwatersheds is obtained.

4 Check the watershed grid's VAT table to see how many cells are there in each subwatershed. If the values are consistently and significantly over 500,000, then reduce the threshold value and repeat steps 2 through 4.

5 Once the satisfactory threshold number is determined, finish the preprocessing tasks (vector processing).

The smaller threshold will result in a larger number of subwatersheds. This will speed up the delineation for a point (as the extracted flow direction grid is smaller), but will increase the time needed to perform all the required preprocessing steps, and will increase the size of the preprocessed themes (but not significantly). Since preprocessing is done only once, this is not a major issue. At some point, reducing the size of the subwatersheds will not return any benefits, as the overhead of grid extraction and polygon addition will become larger than the actual delineation within the subwatershed.

The second aspect of threshold selection concerns coastal and inner drainage basins. Using the current methodology, in order to define an initial subwatershed, a stream link needs to be present for that subwatershed. The stream link is a derivative of the stream, which in turn is defined by the threshold. For a large threshold, it is possible that smaller areas draining directly into the sea or a lake will not be identified (areas that drain less than the threshold value).

Figure 11 shows this issue on a North American HYDRO 1K data set (USGS et al.). The cell size for the DEM is 1 kilometer. The DEM covering the conterminous United States contains 3,184 rows by 4,650 columns, for a total of about 14,800,000 cells. The selected threshold had 10,000 cells corresponding to 10,000 square kilometers (before a stream was identified). The DEM-to-threshold-size ratio is about 1/1,480 (significantly larger than the recommended starting point—resulting in a denser subwatershed theme), producing 411 subwatersheds, ranging in size from 10 to 120,000 cells. The time to perform point delineation is about four seconds.

Figure 11, focusing on the southeastern United States, shows that large portions of the coastal areas have not been predelineated. In these areas, the watershed delineation methodology cannot be used, and the traditional method must be applied. To speed up the delineation in those areas, a hybrid approach to watershed delineation can be implemented (using the same principle of limiting the extent of the grid that needs to be processed):

1 Form a subwatershed theme that contains both the predelineated subwatershed polygons and the polygons for areas that fall under threshold size. Note that by definition the polygons that are left out will be small (less than the threshold), and in general non-contiguous.

2 If the point falls in one of the preprocessed watersheds, perform delineation following the FWD methodology already described.

If the point falls in one of the left-out polygons, extract the flow direction grid and perform the delineation using GIS intrinsic functions. Since we know that these areas are not connected to the upstream drainage area, there is no need to connect the resulting watershed to the premerged upstream polygon, and the delineation is completed (similar to the operations on head basins).

The performance of this method for delineation in coastal areas will be similar to or faster than that for the inner basins since there is no need to add the upstream polygons. The inner basins (lakes) can be processed in the same way.

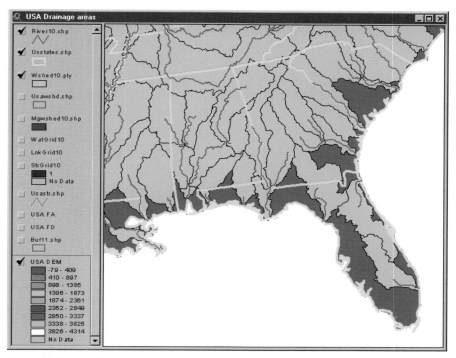

Figure 11. Threshold issues related to initial subwatershed delineation in coastal areas. The darker areas indicate watersheds not included in the set of predelineated subwatersheds.

IMPLEMENTATION OF THE WATERSHED DELINEATION METHODOLOGY

The described methodology was first implemented in the Watershed Delineator (ESRI, 1997), a public-domain ArcView GIS extension (requiring the ArcView Spatial Analyst extension) that ESRI developed for the Texas Natural Resource Conservation Commission for the sole purpose of efficiently delineating watersheds. The extension contains all the necessary tools for FWD methodology implementation, including preprocessing and watershed delineation. The watersheds for points, segments, or polygons can be delineated either interactively or in batch mode where the data of interest are provided to the application as shapefiles.

Currently, ESRI and HEC are using this approach to develop another ArcView GIS extension to support interactive watershed development for hydrologic modeling. Besides watershed delineation, this extension will have tools for topographic characteristic extraction, hydrologic parameter definition, and model input file generation, similar to CRWR-PrePro (Olivera and Maidment, 1999).

From the watershed delineation point of view, the key difference is that the final result of the application is a set of nonoverlapping subwatersheds, of which each one will be modeled as a hydrologic unit. The described watershed delineation methodology is used to identify the area of interest for which the hydrologic modeling will be performed. Once this area is identified, a set of interactive tools are made available for refinement of the initial delineation (merge and split current subwatershed, add subbasin outlet points from a file, and so on). The core delineation functions (point, segment, and polygon) will also be included with this application to provide a full set of delineation tools.

CONCLUSIONS

The FWD methodology described in this paper allows efficient and consistent watershed delineation on DEMs of any size. The speed of delineation can be controlled by the user during the preprocessing stages, and can be on the order of five to ten seconds per watershed on most of today's desktop systems, allowing for a truly interactive operation. The required preprocessing needs to be done only once for a given DEM and can be done at a different location from the one where the actual delineation work will be performed. This allows preprocessing of regional data sets that are then distributed to the end users, who can immediately apply delineation tools to those data sets.

The FWD methodology and the tools developed for its implementation are only effective if a good-quality DEM is available. The quality of a DEM is a function of the scale and the task at hand. The described methodology promotes development of quality data as these issues have to be addressed sooner (during the preprocessing stages) rather than later (during watershed delineation). These issues can then be addressed by a group of experienced analysts, instead of by end users who are not versed in the intricacies of terrain representation using DEM structures.

ACKNOWLEDGMENTS

The Watershed Delineation application that prompted initial development of the FWD methodology was built by ESRI for the Texas Natural Resource Conservation Commission (Austin, Texas) under the WIPS grant from the Environmental Protection Agency.

The current implementation of the methodology is under the CRADA agreement between ESRI and the U.S. Army Corps of Engineers Hydrologic Engineering Center (Davis, California).

REFERENCES

Cederstrand, J. R., and A. Rea. 1995. Watershed boundaries and digital elevation model of Oklahoma derived from 1:100,000-scale digital topographic maps. U.S. Geological Survey Open-File Report 95-727.

Djokic, D., Z. Ye, and A. Miller. 1997. Efficient Watershed Delineation Using ArcView and Spatial Analyst. In *1997 ESRI International User Conference Proceedings,* Environmental Systems Research Institute, Inc., Redlands, California.

ESRI. 1997. *Watershed Delineator Application: User's Manual.* Environmental Systems Research Institute, Inc., Redlands, California.

Garbrecht, J., and L. W. Martz. 1999. Digital Elevation Model Issues in Water Resources Modeling. In *1999 ESRI International User Conference Proceedings* (paper 1 in this book), Environmental Systems Research Institute, Inc., Redlands, California.

Olivera, F., and D. R. Maidment. 1999. GIS Tools for HMS Modeling Support. In *1999 ESRI International User Conference Proceedings* (paper 5 in this book), Environmental Systems Research Institute, Inc., Redlands, California.

Perez, A. 1999. Source Water Protection Project: A Comparison of Watershed Delineation Methods in ARC/INFO and ArcView GIS. In *1999 ESRI International User Conference Proceedings* (paper 3 in this book), Environmental Systems Research Institute, Inc., Redlands, California.

USGS EROS Data Center. HYDRO 1K Elevation Derivative Database. http://edcwww.cr.usgs.gov/landdaac/gtopo30/hydro.

ABOUT THE AUTHORS

Dean Djokic, Ph.D. is a senior applications programmer and consultant at ESRI, where he is focusing on development of custom applications in the area of water resources and water resources modeling. Prior to ESRI, Dr. Djokic worked as a consultant in ElektroProjekt (Zagreb, Croatia) on hydrologic and hydraulic analyses; as a research associate in the area of GIS applications in water resources at the Center for Research in Water Resources, the University of Texas at Austin; and as assistant professor at the University of New South Wales in Sydney, Australia. He holds B.S. and M.S. degrees in civil engineering from the University of Zagreb; the Diploma Hydrology from IHE, Delft; and a Ph.D. from the University of Texas at Austin.

Zichuan Ye, Ph.D. has been an application programmer at ESRI since 1996. Dr. Ye is primarily interested in developing the GIS applications in the areas of water resources modeling, planning, and management as well as GIS applications over the Internet. He holds a B.S. in civil engineering from Fuzhou university, a M.S.E. from IWHR in Beijing, and M.P.A., M.S.E., and Ph.D. degrees from the University of Texas at Austin.

CONTACT THE AUTHORS

Dr. Dean Djokic
Senior Applications Programmer and Consultant
ESRI
380 New York Street
Redlands, CA 92373-8100
Telephone: 909-793-2853, extension 1072
Fax: 909-307-3014
ddjokic@esri.com

Dr. Zichuan Ye
Applications Programmer
ESRI
380 New York Street
Redlands, CA 92373-8100
Telephone: 909-793-2853, extension 1505
Fax: 909-307-3014
zye@esri.com

Paper 5

GIS Tools for HMS Modeling Support

FRANCISCO OLIVERA AND DAVID R. MAIDMENT

CENTER FOR RESEARCH IN WATER RESOURCES

UNIVERSITY OF TEXAS AT AUSTIN

AUSTIN, TEXAS

ABSTRACT

CRWR-PREPRO is a system of ArcView GIS scripts and associated controls, developed to extract topographic, topologic, and hydrologic information from digital spatial data of a hydrologic system, and to prepare ASCII files for the basin and precipitation components of HEC-HMS. These files, when opened by HEC-HMS, automatically create a topologically correct schematic network of subbasins and reaches attributed with hydrologic parameters, and a protocol to relate gage to subbasin precipitation time series. Starting from the DEM, CRWR-PrePro delineates the subbasins and the reach network, calculates parameters for each hydrologic element, determines their interconnectivity, and prepares an input file for HEC-HMS that includes the computed hydrologic parameters. CRWR-PrePro also generates an input file for the precipitation component of HEC-HMS. Two methods of interpolating precipitation records are supported: one to calculate average precipitation at the subbasins based on Thiessen polygons, and another to calculate the routing parameters of precipitation cells for hydrograph determination. Using CRWR-PrePro, the determination of the spatial parameters for HEC-HMS is a simple and automatic process that accelerates the setting up of a hydrologic model and leads to reproducible results.

Rainfall runoff modeling and flood discharge estimation have always been important tasks in hydrologic sciences and engineering. Flood flow estimation, in particular, has been given special attention because of the impact that accurate forecasts have in the management of flood-related emergency programs. Probably more than other concerns in hydrology, estimation of flood discharges is oriented toward saving human lives and protecting people's property.

The Hydrologic Modeling System (HMS), developed by the Hydrologic Engineering Center (HEC) of the U.S. Army Corps of Engineers (USACE), is a software package used to model rainfall runoff processes in a watershed or region, and is a further development of the well-known HEC program HEC-1. For rainfall runoff modeling, HMS requires three input components:

- Basin component, which is a description of the different elements of the hydrologic system (subbasins, channels, junctions, sources, sinks, reservoirs, and diversions) including their hydrologic parameters and topology

- Precipitation component, which is a description—in space and time—of the precipitation event to be modeled, and consists of time series of precipitation at specific points or areas and their relation to the hydrologic elements

- Control component, which defines the time window for the precipitation event and for the calculated flow hydrograph

The first two components depend strongly on spatial factors, so geographic information systems constitute a powerful tool to generate this type of input data.

CRWR-PrePro is a system of ArcView GIS scripts and associated controls developed to extract topographic, topologic, and hydrologic information from digital spatial data of a hydrologic system, and to prepare ASCII files for the basin and precipitation components of HEC-HMS. These files, when opened by HEC-HMS, automatically create a topologically correct schematic network of subbasins and reaches attributed with hydrologic parameters and a protocol to relate gage and subbasin precipitation time series. CRWR-PrePro was developed at the Center for Research in Water Resources (CRWR) of the University of Texas at Austin, and supersedes the former CRWR package HEC-PrePro version 2.0 (Olivera and Maidment, 1998a). Additional capabilities of CRWR-PrePro with respect to HEC-PrePro version 2.0 include the relaxation of the one-to-one subbasin/reach relation that was enforced in the previous version, subsystem extraction by selecting the downstream subbasin polygon, and preparation of an input file for the precipitation component, among others.

HEC-HMS is a flexible program that allows the user to choose among different loss rate, subbasin routing, and baseflow models for the subbasins, as well as different routing methods for the reaches. However, because some of these models and methods depend on hydrologic parameters that cannot be extracted from readily available spatial data, CRWR-PrePro does not estimate parameters for all of the methods supported by HEC-HMS. At the moment, CRWR-PrePro calculates or imports parameters for the following:

- Soil Conservation Service (SCS) curve number method and the initial plus constant loss method for loss rate calculations

- SCS unit hydrograph model for subbasin routing for which the lag time can be calculated with the SCS lag-time formula or as a fraction of the length of the longest channel divided by the flow velocity

- Muskingum method and the lag method for flow routing in the reaches (depending on the reach length)

Using CRWR-PrePro, determining the spatial parameters for HEC-HMS is a simple and automatic process that accelerates setting up a hydrologic model for HEC-HMS and leads to reproducible results.

PREVIOUS WORK

The suitability of raster-based GIS for modeling gravity-driven flow has been previously addressed in the literature (Olivera and Maidment 1998a, Maidment 1992). Consequently, raster-based GIS algorithms for hydrologic analysis have been developed (Jensen and Domingue 1988, Jensen 1991) and included in commercially available GIS software. Functions to delineate reaches and subbasins that use Jensen and Domingue's algorithms are available in ArcView GIS 3.0a Spatial Analyst 1.1 through Avenue requests and also through the Hydrologic Modeling extension to ArcView GIS distributed by ESRI with ArcView Spatial Analyst. Likewise, digital elevation models (DEMs) have been developed for different parts of the world at different resolutions (USGS a, USGS b, USGS c), and further developments of these data have been aimed at improving their spatial resolution. Other spatial data sets such as land use and soil type have also been developed for different parts of the world.

CRWR-PrePro is the synthesis of ArcView GIS applications developed over the last few years at ESRI and the CRWR. The Watershed Delineator extension to ArcView GIS (Djokic et al. 1997, ESRI 1997), developed by the Applications Programming group at ESRI for the Texas Natural Resources Conservation Commission (TNRCC), can be used to delineate watersheds to a point, line segment, or polygon, selected interactively by the user from the map. The Flood Flow Calculator extension to ArcView GIS (Olivera et al. 1997, Olivera and Maidment 1998b), developed at the CRWR for the Texas Department of Transportation (TxDOT), can be used to estimate flood peak flows according to the regional regression equations developed by Asquith and Slade (1997) for Texas. All hydrologic parameters required by these equations, such as drainage area, watershed shape factor, and slope of the longest flow path, are extracted automatically from the spatial data. HECPREPRO (Hellweger and Maidment 1999), developed at the CRWR for the HEC, can be used to establish the topology of the hydrologic elements, and write an input ASCII file readable by HEC-HMS with all this information. CRWR-PrePro combines the terrain analysis capabilities of the Watershed Delineator with the hydrologic parameter calculation capabilities of the Flood Flow Calculator and the topologic analysis capabilities of HECPREPRO to conform a hydrologic modeling tool that prepares—from readily available digital spatial data—the input file for the HEC-HMS basin component. CRWR-PrePro uses code originally developed for the Watershed Delineator, Flood Flow Calculator, and HECPREPRO, although modifications have been made to meet the specific needs of this system.

Additionally, precipitation interpolation methods, developed at the CRWR and HEC, have been included in CRWR-PrePro for determining the HEC-HMS precipitation input component. A Thiessen-polygon-based method (Dugger 1997), developed at the CRWR, can be used to estimate subbasin precipitation as a weighted average of gage precipitation. GridParm (HEC 1996), originally developed at HEC in ARC Macro Language (AML) and rewritten at the CRWR as an Avenue script, can be used to determine parameters of precipitation cells for use with the ModClark subbasin routing method of HEC-HMS. ModClark, a variation of the original Clark unit hydrograph model, has been developed for HEC-HMS as a subbasin routing option suitable to support NEXRAD precipitation data.

METHODOLOGY CRWR-PrePro generates data for the basin and precipitation input components of HEC-HMS.

INPUT DATA FOR THE HEC-HMS BASIN COMPONENT

The process of generating input data for the basin component has been divided into six conceptual modules: (1) raster-based terrain analysis; (2) raster-based subbasin and reach network delineation; (3) vectorization of subbasins and reach segments; (4) computation of hydrologic parameters of subbasins and reaches; (5) extraction of hydrologic subsystem (if necessary); (6) topologic analysis and preparation of the HEC-HMS basin file.

Raster-based terrain analysis

Raster-based terrain analysis for hydrologic purposes uses Jensen and Domingue's (1988) algorithms. By running the FLOWDIRECTION Avenue request, a single downstream cell—out of its eight neighbor cells—is defined for each terrain cell. This downstream cell is selected so that the descent slope from the cell is the steepest. Therefore, a unique path from each cell to the basin outlet is determined. This process produces a reach network, with the shape of a spanning tree, that represents the paths of the watershed flow system. However, because a flow direction cannot be determined for cells that are lower than their surrounding neighbor cells, a process of filling the spurious terrain pits is necessary before determining the flow directions (ESRI 1992). Once the terrain depressions have been filled and the flow directions are known, the drainage area—in units of cells—is calculated with the FLOWACCUMULATION Avenue request. The flow accumulation grid stores the number of cells located upstream of each cell (the cell itself is not counted) and, if multiplied by the cell area, equals the drainage area. Figure 1 shows an example of how the FLOWDIRECTION and FLOWACCUMULATION requests work when applied to a DEM.

32	64	128
16	✳	1
8	4	2

Flow direction codes

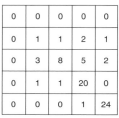

78	72	69	71	58
74	67	56	49	46
69	53	44	37	38
64	58	55	22	31
68	61	47	21	16

Digital elevation model (DEM)

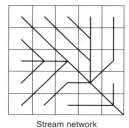

2	2	2	4	4
2	2	2	4	4
1	1	2	4	8
128	128	1	2	4
128	128	1	1	4

Flow direction grid

0	0	0	0	0
0	1	1	2	1
0	3	8	5	2
0	1	1	20	0
0	0	0	1	24

Flow accumulation grid

Stream network

Figure 1. Raster-based functions for terrain analysis for hydrologic purposes.

Raster-based subbasin and reach network delineation

The DEM cells that form the reaches are defined as the union of two sets of grid cells. The first set consists of all cells whose flow accumulation is greater than a user-defined threshold value. This set identifies the reaches with the largest drainage area but not necessarily with the largest flow, because flow depends on other variables that are not related exclusively to topography. The second set is defined interactively by the user by clicking on a certain point on the map, which results in an automatic selection of all downstream cells. This capability allows the user to select a particular reach, which might have a small drainage area (low flow accumulation), without having to lower the threshold value for the entire system and defining unnecessarily a denser reach network. After the reach cells have been defined, a unique identification number or grid code is assigned to each reach segment. Figure 2 shows threshold-based and user-defined reaches, as well as their corresponding reach segments.

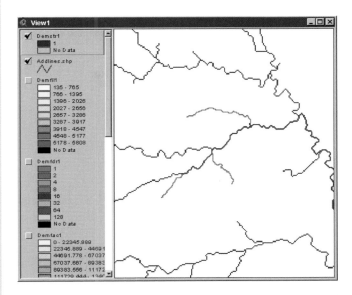

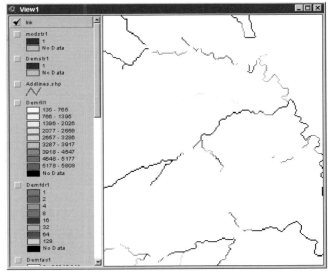

Figure 2. Reach network delineation. In the upper figure, blue cells correspond to drainage areas greater than 3,000 grid cells, whereas red cells are defined interactively. In the lower figure, each reach segment has been identified with a different grid code and displayed with a different color.

Subbasin outlets are also defined as the union of two sets of grid cells. The first set, based on the reach network, consists of all cells located just upstream of the junctions. Consequently, at a junction, two outlet cells are identified, one for each of the upstream branches.

The system outlet is also identified as a subbasin outlet. Since these outlets are the most downstream cells of the reach segments, their identification number or grid code is the same as their corresponding reach segment. The second set is defined interactively by the user by clicking on any cell of the reach network, such as those associated with flow gages, reservoirs, or other water control points. The identification number or grid code of each new interactively defined outlet is obtained by adding one to the highest grid code value available.

Reach segments containing interactively defined outlets are subdivided at the clicked cells, so that the new segments—upstream of the new outlets—are assigned the same grid code as their corresponding new outlet. Outlets associated with reservoirs can be identified so that HEC-HMS recognizes them as both reservoirs and subbasin outlets. Figure 3 shows threshold-based and user-defined outlets, as well as the corresponding reach segments.

The watershed Avenue request is then used to delineate the areas draining to each subbasin outlet. Subbasins are assigned the same identification number, or grid code, as their corresponding outlet and reach segment. Figure 4 shows the delineated subbasins and reach network in raster format.

At this point, a one-to-one relation between reach segments and subbasins is maintained because a unique subbasin outlet has been identified for each reach segment. Due to this one-to-one relation, subbasin and reach segment grid codes are equal.

Vectorization of subbasins and reach segments

Because HEC-HMS applies lumped models within each hydrologic element, hydrologic parameters have to be calculated for the subbasins and reach segments, and not for the individual grid cells. After the reach segments and their corresponding drainage areas have been delineated in the raster domain, a vectorization process is performed using raster-to-vector conversion functions. This process consists of creating a polyline feature data set of reaches and a polygon feature data set of subbasins. The grid code values are transferred to the attribute tables of the feature data sets, thus preserving a way to directly link subbasins and reaches. Further vector processing (i.e., merging of dangling polygons) might be necessary to ensure that each subbasin is represented by a single polygon, and that the one-to-one reach/subbasin relation established in the raster domain is preserved in the vector domain (Olivera et al. 1998a). Figure 5 shows the delineated subbasins and reach network after vectorization.

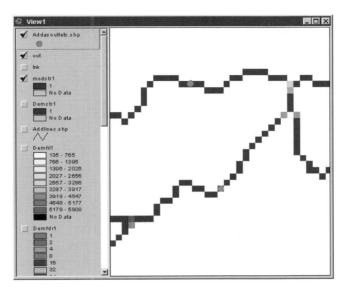

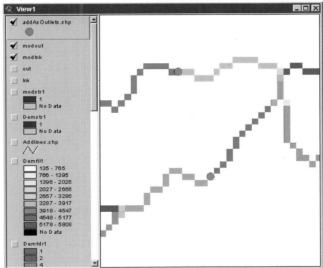

Figure 3. User-defined subbasin outlets. In the upper figure, blue cells represent the stream network in raster format; color cells, the subbasin outlets located just upstream of the junctions; and red dots, interactively defined outlets. In the lower figure, each stream segment is displayed in a different color, and segments containing red dots have been subdivided into two or more segments.

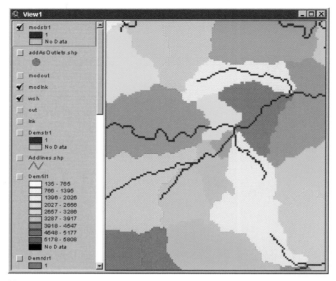

Figure 4. Delineated subbasins and reach network in raster format.

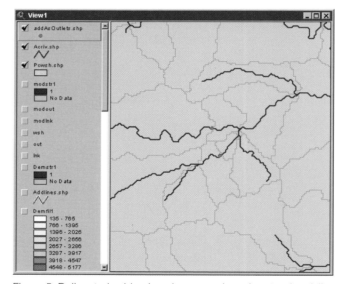

Figure 5. Delineated subbasin polygons and reach network polylines after vectorization.

The one-to-one reach/subbasin relation, though, can be relaxed by merging adjacent subbasin polygons, so that a subbasin contains more than one reach. In such a case, a new field is necessary in the attribute table of the reaches to account for the grid code of the subbasin in which the reach is located after merging polygons. For merging two subbasins, the polygons have to share the same outlet or drain one toward the other. Figure 6 shows the merging of two subbasins that share the same outlet, as well as the attribute tables of the subbasin and reach network data sets before and after the merging.

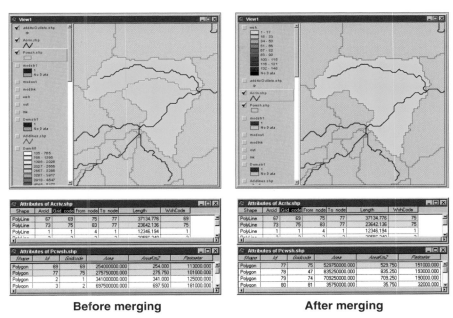

Before merging   **After merging**

Figure 6. Subbasin polygons and attribute tables before and after merging polygons. In this case, the two merged polygons share a common outlet.

CRWR-PrePro also has the capability of identifying, for each subbasin polygon, all the subbasin polygons located upstream of it, so that they can be easily retrieved when delineating a watershed from a point.

Computation of hydrologic parameters of subbasins and reaches

The subbasin parameters calculated by CRWR-PrePro are area, lag time, and average curve number. Other parameters needed for estimating the lag time, such as length and slope of the longest flow path, are also calculated and stored in the subbasin attribute table. Depending on the algorithm used to calculate the lag time, it might depend entirely on spatial data (i.e., DEM, land use, and soils) or it might require additional externally supplied input. Depending on the method selected, the average curve number can be used to calculate the subbasin lag time and the subbasin loss rate. Figure 7 shows the attribute table of the subbasins data set with the calculated hydro-logic parameters appended.

Shape	Id	Gridcode	Area	AreaKm2	Perimeter	LngFlwPth	Slope	Baseflow	Transform	CurveNum	LagTime
Polygon	25	28	486750000.000	486.750	198000.000	51334.5078	0.0035	None	SCS	85.6119	1615.3280
Polygon	40	29	820500000.000	820.500	192000.000	70183.7344	0.0026	None	SCS	81.0727	2804.3778
Polygon	24	30	418500000.000	418.500	106000.000	35727.9258	0.0038	None	SCS	82.4433	1292.8682
Polygon	33	31	509250000.000	509.250	155000.000	46041.9375	0.0040	None	SCS	84.0697	1461.6324
Polygon	31	32	85000000.000	85.000	54000.000	18985.2813	0.0023	None	SCS	85.2711	910.0928
Polygon	30	33	64750000.000	64.750	53000.000	16363.9600	0.0030	None	SCS	85.5036	701.7632
Polygon	29	34	291000000.000	291.000	97000.000	31263.4609	0.0044	None	SCS	83.3809	1046.6376
Polygon	41	35	625500000.000	625.500	171000.000	59526.8906	0.0031	None	SCS	88.5953	1729.0834

Figure 7. Subbasin attribute table showing the calculated and appended fields: area in km^2 (AreaKm2), length of longest flow-path (LngFlwPth), slope of longest flow-path (Slope), baseflow (Baseflow), subbasin routing method (Transform), average curve number (CurveNum), and subbasin lag time (LagTime).

The subbasin area is calculated automatically in the process of vec-torizing the subbasin polygons.

The subbasin lag time is calculated with either of the following formulas:

$$t_p = \max\left(\frac{L_w^{0.8}[(1000/CN)-9]^{0.7}}{31.67 S^{0.5}}, 3.5\Delta t\right) \tag{1}$$

$$t_p = \max\left(0.6\frac{0.3048 L_w}{60 v_w}, 3.5\Delta t\right) \tag{2}$$

where t_p (minutes) is the subbasin lag time measured from the centroid of the hyetograph to the peak time of the hydrograph, L_w (feet) is the length of the longest flow-path, S (%) is the slope of the longest flow-path, CN is the average curve number in the subbasin, Δt (min) is the analysis time-step, and v_w (m/s) is a representative velocity of the longest flow-path. In equation 1, the first term in the parentheses corresponds to the lag time according to the SCS (1972), whereas the second term is a minimum lag-time value required by HEC-HMS (HEC 1990). In equation 2, the first term corresponds to the lag time defined as 60 percent of the subbasin time of concentration, and again the second term is a minimum lag-time value required by HEC-HMS. Olivera et al. (1988a) present in detail the methodology to calculate L_w and S from DEM data. Values of v_w cannot be estimated from spatial data and have to be supplied by the user.

CN is calculated as the average of the curve number values within the subbasin polygon. A curve number grid is calculated using land-use data described by Anderson land-use codes, percentage of hydrologic soil group (A, B, C, and D) according to STATSGO soils data, and a lookup table that relates land use and soil group with curve numbers (Smith 1995).

Loss rate in the subbasins can be calculated with either of the following methods: the SCS curve number method for which the average curve number is calculated, or the initial plus constant loss rate method for which the initial and constant rate values have to be supplied by the user. It is likely that in the near future it will be possible to establish a relation between terrain properties and loss rate parameters.

The reach parameters determined by CRWR-PrePro are the length, the routing method (either Muskingum or pure lag), the Muskingum K and the number of subreaches into which the reach is subdivided in case Muskingum is used for routing, and the flow time in case pure lag is used for routing. Other reach parameters like the flow velocity and the Muskingum X cannot be computed from spatial data and must be calculated externally and supplied as input. Figure 8 shows the attribute table of the reach network data set with the calculated hydrologic parameters appended.

Shape	Arcid	Grid_code	From node	To node	Length	WshCode	StreamVel	MuskX	Route	MuskK	NumReachN	LagTime
PolyLine	1	1	4	1	12346.194	1	1.0	0.2	Muskingum	3.4295	2	0.0000
PolyLine	2	2	3	2	28556.349	2	1.0	0.2	Muskingum	7.9323	3	0.0000
PolyLine	3	3	5	12	22788.582	3	1.0	0.2	Muskingum	6.3302	3	0.0000
PolyLine	4	8	11	12	1560.660	8	1.0	0.2	Lag	0.0000	0	26.0110
PolyLine	5	4	8	15	20235.281	4	1.0	0.2	Muskingum	5.6209	2	0.0000
PolyLine	6	12	16	15	4457.107	12	1.0	0.2	Muskingum	1.2381	1	0.0000
PolyLine	7	5	7	17	52412.951	5	1.0	0.2	Muskingum	14.5592	5	0.0000
PolyLine	8	9	12	17	8414.214	9	1.0	0.2	Muskingum	2.3373	1	0.0000
PolyLine	9	15	18	19	750.000	15	1.0	0.2	Lag	0.0000	0	12.5000
PolyLine	10	18	21	22	1060.660	19	1.0	0.2	Lag	0.0000	0	17.6777

Figure 8. Reach attribute table showing the calculated and appended fields: reach velocity (StreamVel), Muskingum X (MuskX), reach routing method (Route), reach flow time in hours or Muskingum K (MuskK), number of subreaches (NumReachN), reach flow time in minutes or lag time (LagTime).

The reach length L (m) is determined automatically in the process of reach vectorization.

The Muskingum method is used for routing in reaches long enough not to present numerical instability problems. In short reaches, in which the flow time is shorter than the time-step, the pure lag method is used. In very long reaches, again to avoid numerical instability, reaches are subdivided into shorter equal-length subreaches, so that the flow time in each of them satisfies the condition 2 X k < Δt < k (HEC 1990), where X is a parameter of the Muskingum method and k (min) is the flow time in the subreach. Since the flow time in the subreaches is equal to k = 60 K / n = (L / 60 v) / n, where K (hrs) is the flow time in the reach, v (m/s) is the reach flow velocity, and n (an integer value greater than zero) is the number of subreaches, then it follows 2 X ((L / 60 v) / Δt) < n < ((L / 60 v) / Δt). Moreover, because n should be at least equal to 1, L / 60 v should be greater than Δt, otherwise the pure lag method must be used as mentioned above.

Thus, the minimum number of subreaches into which the reach should be subdivided is given by

$$n = int \left(2 X \frac{L/60v}{\Delta t} \right) + 1 \qquad (3)$$

whereas the maximum number of subreaches is given by

$$n = int \left(\frac{L/60v}{\Delta t} \right) \qquad (4)$$

where int takes the integer part of the argument (int does not round the number). To avoid unnecessary computations, the number of subreaches is taken as the minimum value given by equation 3.

K (hrs) for the Muskingum method is equal to L/3600 v, and the lag time (min) for the pure lag method is equal to L/60 v.

At present, CRWR-PrePro supports only digital spatial data in horizontal meters and DEM elevations in meters. As well, it generates parameters for use with the HEC-HMS SI units option only.

Extraction of hydrologic subsystem
Extraction of a hydrologic subsystem consists of detaching from the overall study area a set of subbasin polygons and corresponding reach polylines for further hydrologic analysis with HEC-HMS.

Subsystems can be defined either by manually selecting the subbasin polygons or by manually selecting the most downstream subbasin polygon (and automatically selecting the subbasin polygons of its contributing drainage area).

The first method is more flexible, although more tedious to implement. It has no restriction on the polygons that can be selected, and it supports the use of inlets ("sources," according to the HEC-HMS terminology) to represent areas draining to the subsystem. Reach polylines contained within the selected polygons—as well as those draining toward them—are selected automatically. Reach polylines draining toward the selected polygons are used to identify the subsystem inlets. Figure 9 shows a subsystem extraction when the polygons are manually selected.

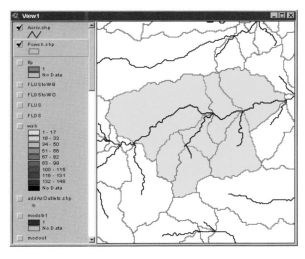

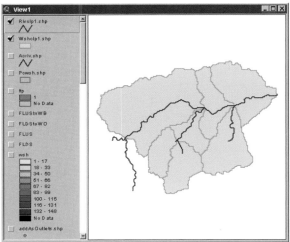

Figure 9. Subsystem extraction by selecting all the relevant subbasin polygons. Upstream reaches are selected to help identify system sources.

The second method is less flexible, but easier to implement. After manually selecting the downstream subbasin polygon, it automatically identifies and selects all the subbasin polygons located upstream, and consequently does not support the use of inlets. Reach polylines contained within the selected polygons are selected automatically. This method is convenient when dealing with a significant number of polygons in the study area. Figure 10 shows a subsystem extraction when only the most downstream polygon is manually selected.

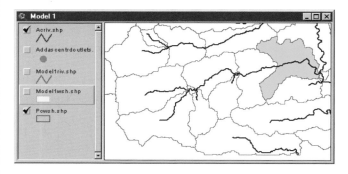

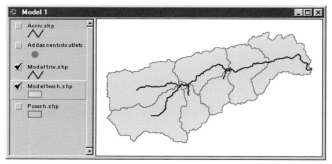

Figure 10. Subsystem extraction by selecting the downstream subbasin polygon manually and selecting all other upstream polygons automatically.

Topologic analysis and preparation of HEC-HMS basin file

Establishing the topology of the hydrologic system consists of determining the element located downstream of each element. Since the HEC-HMS hydrologic schematic allows only one downstream element, no ambiguity is introduced in this process. After establishing the system topology based on the subbasin and reach data sets, an ASCII file—readable by HEC-HMS—is used to record the type (i.e., subbasin, reach, source, sink, reservoir, or junction), hydrologic parameters, and downstream element of each hydrologic element of the system. A background map file—also readable by HEC-HMS—

is used to graphically represent subbasins and reaches, and ease the identification of hydrologic elements. These files constitute the input to the basin component of HEC-HMS. Figure 11 shows three different sections of the basin file.

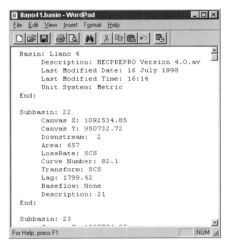
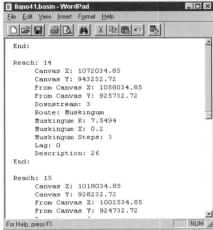
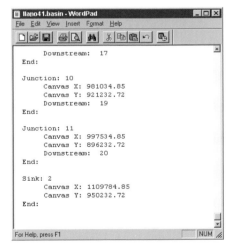

Figure 11. HEC-HMS basin file in ASCII format. Hydrologic parameters calculated in GIS and stored in the attribute tables are transferred to the basin file.

This basin file, when opened with HEC-HMS, generates a topologically correct schematic network of hydrologic elements and displays it in the HEC-HMS Schematic window together with the background map. Figure 12 shows a detail of an HEC-HMS schematic and the corresponding sections of the basin file used to build it.

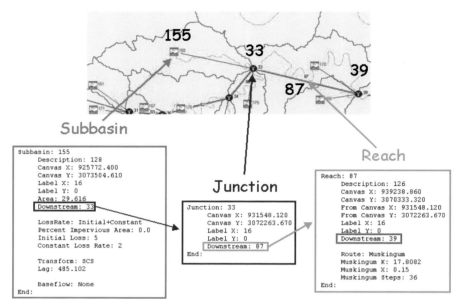

Figure 12. HMS schematic of the hydrologic system constructed from the basin file.

Figure 13 shows the HEC-HMS Schematic window after opening the basin file.

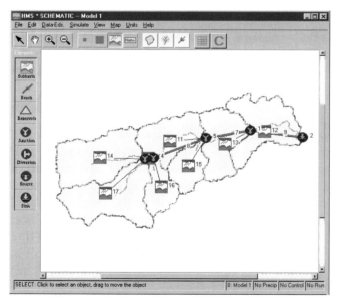

Figure 13. HEC-HMS display of the schematic of the hydrologic system as described by basin input component.

The process of generating input data for the precipitation component consists of calculating precipitation time series for subbasin polygons from precipitation time series at precipitation gages or precipitation cells (i.e., NEXRAD cells). At present, two precipitation methods are supported by CRWR-PrePro: user-specified gage weighting and Grid-Parm. Automatic determination of user-specified gage weights for precipitation interpolation, based on Thiessen and subbasin polygon data sets, was developed by Dugger (1997). Precipitation time series at subbasins are estimated as an area-weighted average of precipitation time series at gages. GridParm (HEC 1996) was originally developed at HEC in ARC Macro Language (AML) and rewritten at CRWR as an Avenue script. GridParm (short for grid cell parameters) is used to determine parameters of precipitation cells for use with the Mod-Clark subbasin routing method of HEC-HMS.

User-specified gage weighting

Subbasin precipitation time series are calculated as the weighted average of gage precipitation time series. For this purpose, a set of weights that capture the relative importance of the precipitation at each gage on the precipitation of each subbasin is calculated. Precipitation time series at the gages are stored in data storage system (DSS) format (HEC 1995). Figure 14 shows precipitation data in text format.

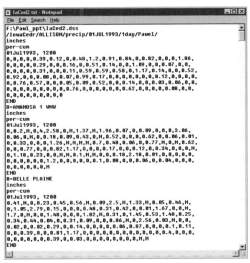

Figure 14. Precipitation data in text format before using HEC-DSS to convert it into DSS format.

Given a set of points that represent gages for which precipitation time series are known, Thiessen polygons are used to establish the area of influence of each precipitation gage. Thiessen polygons are constructed by drawing perpendiculars at the midpoints of the segments that connect the gages, so that all points within a polygon are closer to the polygon gage than to any other gage. By intersecting the Thiessen with the subbasin polygons, a new set of smaller polygons is defined in such a way that each new polygon is related to one Thiessen polygon only and one subbasin polygon only. Figure 15 shows the polygons resulting from the intersection of Thiessen polygons and subbasin polygons.

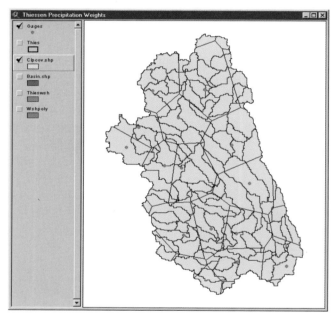

Figure 15. Intersection of subbasin polygons with Thiessen polygons. Red dots indicate precipitation drainage gages.

The ratio of the area of a new polygon to the area of its corresponding subbasin polygon represents the weight of the gage for the subbasin. This can also be expressed as

$$w_{i,j} = \frac{A_{i,j}}{S_j}$$

where $A_{i,j}$ is the area of the polygon generated by intersecting subbasin j with the Thiessen polygon of gage i; S_j is the area of subbasin j; and $w_{i,j}$ is the weight of gage i for subbasin j. The sum of the weights of a subbasin should add up to one. In figure 16, the three selected polygons originally formed a complete subbasin, and their weight is proportional to their area.

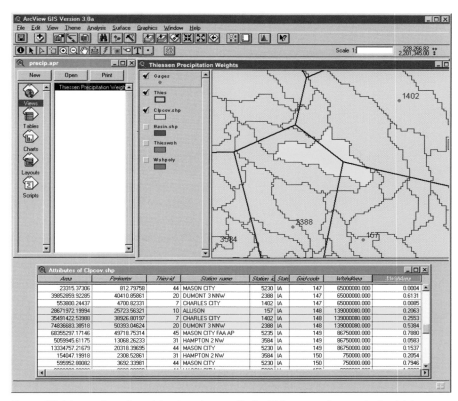

Figure 16. Intersection of a subbasin polygon with the Thiessen polygons. In the table below, the area of each of the yellow polygons is stored in the field Area, the subbasin area in the field WtshdArea, and the weights (the ratio of these two areas) in the field %WshArea.

After establishing the weight values based on the subbasin and Thiessen polygon data sets, an ASCII file is used to record the gage and subbasin information. The gage information consists of the gage name, location, type (i.e., incremental or cumulative), and reference to the precipitation time series in the DSS file. The subbasin information consists of the subbasin name or identification code, and the name of each gage with its corresponding weight. Figure 17 shows the HEC-HMS precipitation weights text file.

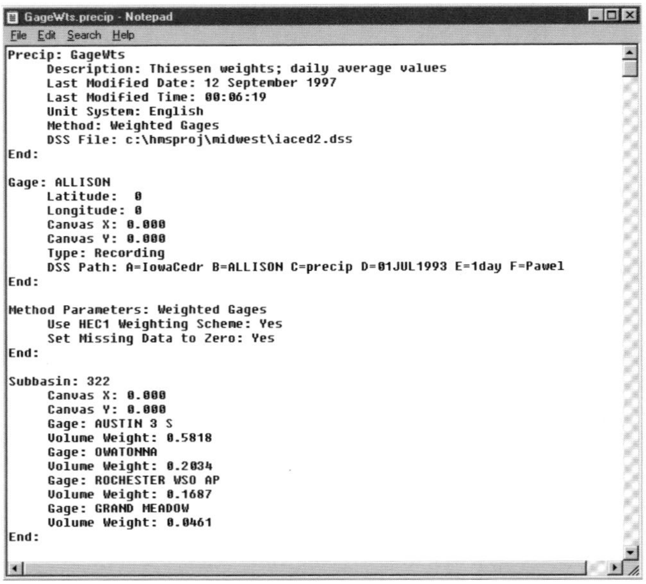

```
Precip: GageWts
     Description: Thiessen weights; daily average values
     Last Modified Date: 12 September 1997
     Last Modified Time: 00:06:19
     Unit System: English
     Method: Weighted Gages
     DSS File: c:\hmsproj\midwest\iaced2.dss
End:

Gage: ALLISON
     Latitude:  0
     Longitude: 0
     Canvas X: 0.000
     Canvas Y: 0.000
     Type: Recording
     DSS Path: A=IowaCedr B=ALLISON C=precip D=01JUL1993 E=1day F=Pawel
End:

Method Parameters: Weighted Gages
     Use HEC1 Weighting Scheme: Yes
     Set Missing Data to Zero: Yes
End:

Subbasin: 322
     Canvas X: 0.000
     Canvas Y: 0.000
     Gage: AUSTIN 3 S
     Volume Weight: 0.5818
     Gage: OWATONNA
     Volume Weight: 0.2034
     Gage: ROCHESTER WSO AP
     Volume Weight: 0.1687
     Gage: GRAND MEADOW
     Volume Weight: 0.0461
End:
```

Figure 17. HEC-HMS precipitation weights file in ASCII format. Gage weights calculated in GIS and stored in the attribute tables are transferred to the precipitation weights file.

Finally, the subbasin precipitation time series are calculated by HEC-HMS as

$$P_j(t) = \sum_i p_i(t) w_{i,j}$$

where $p_i(t)$ is the precipitation time series at gage i, and $P_j(t)$ is the precipitation time series at subbasin j.

GridParm

GridParm (short for grid cell parameters) is used to determine parameters of precipitation cells for use with the ModClark subbasin routing method of HEC-HMS. ModClark, a variation of the Clark unit-hydrograph model (Clark 1945), has been developed by HEC for HEC-HMS as a subbasin routing option to support NEXRAD precipitation data. In the ModClark method, the drainage area is subdivided into elementary cells in which precipitation is uniform and known, and the hydrograph is calculated as the sum of the contribution of each cell (or fraction of cell) within the subbasin. Although cells are usually rectangular since the method was developed to support NEXRAD precipitation data, no restriction on the cell shape exists. Calculation of the routing parameters, area, and flow distance to the outlet, to track the water from the precipitation cell to the outlet, is done by GridParm.

To calculate these parameters, three data sets are defined: (1) subbasin polygons, (2) precipitation cell polygons, and (3) flow length downstream to the subbasin outlet grid. Subbasin polygons are intersected with precipitation cell polygons, yielding a new set of polygons that are complete cells if they were completely within a subbasin, or fractions of cells if they were partially contained by two or more subbasins. Each of these new polygons is called GridCell, and is related to one subbasin polygon only. The average distance from the GridCell to the subbasin outlet is calculated as the mean of the flow-length grid values within the GridCell. Figure 18 shows the intersection of precipitation cells with subbasin polygons.

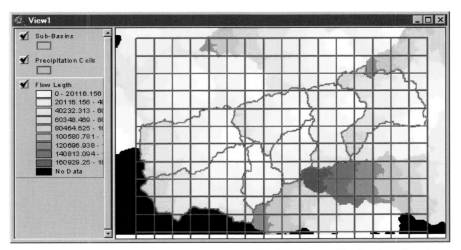

Figure 18. Intersection of precipitation cells with subbasin polygons. The flow length to the subbasin outlet grid is displayed as background.

After establishing the GridCell parameters, an ASCII file is used to record the subbasin and GridCell information. The subbasin information consists of the subbasin name. The GridCell information consists of the location, area, and distance to the subbasin outlet. Figure 19 shows the text file with GridCell parameters as prepared by GridParm.

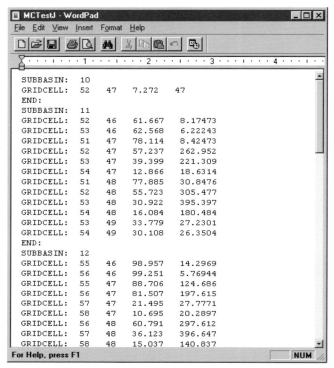

Figure 19. ASCII file with precipitation cell parameters for use with the ModClark subbasin routing method.

CONCLUSIONS

A connection between GIS data sets describing a hydrologic system and HEC's Hydrologic Modeling System (HEC-HMS) has been developed and called CRWR-PrePro.

CRWR-PrePro extracts topographic, topologic, and hydrologic information from digital spatial data, and prepares an input file for the basin component of HEC-HMS, which, when opened, automatically creates a topologically correct schematic network of subbasins and reaches, and attributes each element with selected hydrologic parameters. CRWR-PrePro also generates an input file for the precipitation component of HEC-HMS. Two methods to interpolate precipitation records are supported: Thiessen polygons to calculate average precipitation at the subbasins, and GridParm to calculate routing parameters of the precipitation cells for use with the ModClark subbasin routing method.

At the moment, CRWR-PrePro calculates or imports parameters for the Soil Conservation Service (SCS) curve number method and the initial plus constant loss method for loss rate calculations; the SCS unit hydrograph model for subbasin routing for which the lag time can be calculated with the SCS lag-time formula or as a fraction of the length of the longest channel divided by the flow velocity; and the Muskingum method and the lag method for flow routing in the reaches (depending on the reach length).

Using CRWR-PrePro, determining physical parameters for HEC-HMS is a simple and automatic process that accelerates the setting up of a hydrologic model and leads to reproducible results.

ACKNOWLEDGMENTS

Development of CRWR-PrePro has been funded by the Texas Department of Transportation (TxDOT) and the Hydrologic Engineering Center (HEC) of the U.S. Army Corps of Engineers. The contribution of Joaquim Pinto da Costa at the Instituto da Agua (INAg) of Portugal is appreciated.

REFERENCES

Asquith, W., and R. Slade. 1997. Regional Equations for Estimation of Peak Stream-flow Frequency for Natural Basins in Texas. USGS Water-Resources Investigations Report 96-4307, Austin, Texas.

Clark, C. O. 1945. Storage and the Unit Hydrograph. *Trans. American Society of Civil Engineers* 110:1419–88.

Djokic, D., Z. Ye, and A. Miller. 1997. Efficient Watershed Delineation Using ArcView and Spatial Analyst. In *1997 ESRI International User Conference Proceedings,* Environmental Systems Research Institute, Inc., Redlands, California.

Dugger, A. 1997. Linking GIS with the Hydrologic Modeling System: An Investigation of the Midwest Flood of 1993. Masters Report, Department of Civil Engineering, University of Texas at Austin, Texas.

ESRI. 1992. Cell-based Modeling with GRID 6.1: Supplement—Hydrologic and Distance Modeling Tools. Environmental Systems Research Institute, Inc., Redlands, California.

ESRI. 1997. *Watershed Delineator Application—User's Manual.* Environmental Systems Research Institute, Inc., Redlands, California.

HEC. 1990. *HEC-1—Flood Hydrograph Package—User's Manual.* Hydrologic Engineering Center, U.S. Army Corps of Engineers, Davis, California.

HEC. 1995. *HEC-DSS—User's Guide and Utility Manuals.* Hydrologic Engineering Center, U.S. Army Corps of Engineers, Davis, California.

HEC. 1996. *GridParm—Procedures for Deriving Grid Cell Parameters for the Mod-Clark Rainfall-Runoff Model—User's Manual.* Hydrologic Engineering Center, U.S. Army Corps of Engineers, Davis, CA.

Hellweger, F., and D. R. Maidment. 1999. Definition and Connection of Hydrologic Elements Using Geographic Data. ASCE, *Journal of Hydrologic Engineering* Vol. 4, No. 1.

Jensen, S. K., and J. O. Domingue. 1988. Extracting Topographic Structure from Digital Elevation Data for Geographic Information System Analysis. *Photogrammetric Engineering and Remote Sensing* 54(11).

Jensen, S. K. 1991. Applications of Hydrologic Information Automatically Extracted from Digital Elevation Models. *Hydrological Processes* 5(1).

Maidment, D. R. 1992. Grid-based Computation of Runoff: A Preliminary Assessment. Hydrologic Engineering Center, U.S. Army Corps of Engineers, Davis, California.

Olivera, F., J. Bao, and D. R. Maidment. 1997. Geographic Information System for Hydrologic Data Development for Design of Highway Drainage Facilities. Research Report 1738-3, Center for Transportation Research, University of Texas at Austin, Texas.

Olivera, F., and D. R. Maidment. 1998a. HEC-PrePro v. 2.0: An ArcView Pre-Processor for HEC's Hydrologic Modeling System. In *1998 ESRI International User Conference Proceedings,* Environmental Systems Research Institute, Inc., Redlands, California.

Olivera, F., and D. R. Maidment. 1998b. GIS for Hydrologic Data Development for Design of Highway Drainage Facilities. Transportation Research Record 1625, pp. 131–138, Transportation Research Board, Washington, D.C.

Smith, P. 1995. Hydrologic Data Development System. Masters Thesis, Department of Civil Engineering, University of Texas at Austin, Texas.

Soil Conservation Service. 1972. *National Engineering Handbook, Section 4: Hydrology.* U.S. Department of Agriculture, Washington, D.C.

USGS a. 7.5-Minute Digital Elevation Model Data. http://edcwww.cr.usgs.gov/glis/hyper/guide/7_min_dem as of July 6, 1999.

USGS b. 1-Degree Digital Elevation Models. http://edcwww.cr.usgs.gov/glis/hyper/guide/1_dgr_dem as of July 6, 1999.

USGS c. Global 30 arc-second Elevation Data Set. http://edcwww.cr.usgs.gov/landdaac/gtopo30/gtopo30.html as of July 6, 1999.

ABOUT THE AUTHORS

Francisco Olivera, Ph.D. is a research scientist at the Center for Research in Water Resources at the University of Texas at Austin. Dr. Olivera has conducted and participated in research projects for the Environmental Protection Agency, the Texas Department of Transportation, the Lower Colorado River Authority, the City of Austin, the National Science Foundation, and the Food and Agricultural Organization of United Nations in the areas of hydrology, hydraulics, and the application of GIS to water resources engineering. He has also served as a consultant to the Food and Agricultural Organization of the United Nations, and to the French, Portuguese, and Moroccan governments. Dr. Olivera has published papers in technical journals and presented at national and international conferences. A member of the American Society of Civil Engineers and the American Geophysical Union, he holds a doctoral degree in civil engineering from the University of Texas at Austin, a master of science in hydraulic engineering from the International Institute for Hydraulic and Environmental Engineering at Delft, Holland, and the title of Civil Engineer from the Catholic University of Peru.

David R. Maidment, Ph.D. is the Ashley H. Priddy Centennial Professor of Engineering and director of the Center for Research in Water Resources at the University of Texas at Austin, where he has been on the faculty since 1981. Prior to joining the University of Texas faculty, Dr. Maidment was a research scientist at the Ministry of Works and Development in New Zealand and at the International Institute for Applied Systems Analysis in Vienna, Austria, and was also a Visiting Assistant Professor at Texas A&M University. He is a specialist in surface water hydrology, in particular in the application of GIS to hydrology. He and his research team are currently working on projects applying GIS to floodplain mapping, water quality modeling, water resources assessment, hydrologic simulation, surface water-groundwater interaction, and global hydrology. Each year since 1991, Dr. Maidment has taught a semester-long GIS in Water Resources course. During the fall semester of 1998, he converted this into a virtual course using the Internet to deliver information to off-campus students. ESRI is in the process of providing a summarized version of this course on its virtual campus. He is also coauthor of the text *Applied Hydrology* (McGraw-Hill, 1988), and editor-in-chief of the *Handbook of Hydrology* (McGraw-Hill, 1993). From 1992 to 1995 he was editor of the *Journal of Hydrology,* and is currently an associate editor of that journal and of the *Journal of Hydrologic Engineering.* He earned his master of science and doctoral degrees in civil engineering at the University of Illinois at Urbana–Champaign.

CONTACT THE AUTHORS

Dr. Francisco Olivera
Research Scientist
University of Texas at Austin, Center for Research in Water Resources
J. J. Pickle Research Campus Number 119
Austin, TX 78712
Telephone: 512-471-0570
Fax: 512-471-0072
folivera@mail.utexas.edu

Dr. David R. Maidment
Professor of Civil Engineering
University of Texas at Austin, Center for Research in Water Resources
J. J. Pickle Research Campus Number 119
Austin, TX 78712
Telephone: 512-471-0065
Fax: 512-471-0072
maidment@mail.utexas.edu

Paper 6

Hydrologic Model of the Buffalo Bayou Using GIS

JAMES H. DOAN

U.S. ARMY CORPS OF ENGINEERS — HYDROLOGIC ENGINEERING CENTER

DAVIS, CALIFORNIA

ABSTRACT

THE HYDROLOGIC MODEL of the Buffalo Bayou watershed was simulated using the Hydrologic Modeling System with inputs derived from the geographic information system. The Buffalo Bayou watershed covers most of the Houston metropolitan area in Texas. The watersheds and streams were delineated from USGS digital elevation models (DEMs) at 30-meter cell resolution and stream data from USGS digital line graphs (DLGs) and EPA river reach files (RF1). Physical watershed parameters were extracted from the CRWR-PrePro program to support hydrologic parameters computation. The model uses ModClark runoff transformation with grid-based NEXRAD radar rainfall.

INTRODUCTION

PURPOSE

The first recorded flood in 1929 in the Buffalo Bayou watershed devastated Houston, Texas. Since then, other floods followed, with similar vigor and intensity. Local governments have been actively dealing with floods by building dams and reservoirs to regulate flows to downstream Buffalo Bayou Creek. With the low-relief terrain in the watershed, local governments have also improved the conveyance capacity of streams by straightening reaches, lining channels, and installing hydraulic structures. Although these flood control measures have brought many benefits, the existing flood control infrastructure has been increasingly burdened due to significant urbanization in recent years. Coupling the tremendous growth in Houston with the potential for a large rainfall event similar in magnitude to the 43 inches in a 24-hour period recorded at nearby Alvin, Texas, the rising concerns over flooding are warranted. To guard against future flooding, the Galveston District of the U.S. Army Corps of Engineers has funded this project as a first step toward achieving the long-term goal of developing a Water Control Data System (WCDS). The WCDS will offer real-time monitoring, active warning, and forecasting of floods in the region as shown in figure 1.

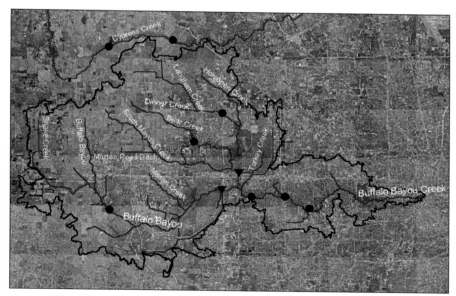

Figure 1. Buffalo Bayou watershed, Houston, Texas.

The Buffalo Bayou project meets many of the long-term objectives for developing a WCDS. The first objective is to develop a hydrologic model using the Hydrologic Modeling System (HMS) for the study area upstream of the Piney Point streamflow gage located on the Buffalo Bayou Creek downstream of the dams. The HMS model of the Buffalo Bayou watershed will employ a quasidistributed rainfall-to-runoff transformation procedure called Modified Clark (Mod-Clark) and grid-based NEXRAD radar rainfall in the Standard Hydrologic Grid (SHG) format to model the watershed response to the storm of October 16–18, 1994. The second objective is to develop, maintain, and use the GIS spatial database of thematic layers to generate inputs for HMS, visualize data, and document watershed conditions.

SCOPE OF WORK

The Galveston District of USACE contracted HEC to develop a hydrologic model using HEC-HMS. The HEC-HMS model will include a basin, precipitation, and control specifications model in accordance with the following guidelines.

Basin model

- The study area includes the Buffalo Bayou watershed upstream of the streamflow gage at Piney Point, including the drainage area tributary to the Addicks and Barker reservoirs. The study area of about 341 square miles is located west of Houston, Texas.

- The GIS spatial database contains spatial data sets to represent the terrain of the study area. At a minimum, the digital elevation models (DEMs) and digital line graphs (DLGs) or river reach files (RF1) shall be included in order to represent the terrain and major streams in the study area.

- The GIS methodologies will be used to delineate subbasins and streams at major stream confluences and specified gage locations. The basin model generated in ASCII format will include proper hydrologic elements with connectivity to represent the drainage patterns.

- Basic topographic parameters, such as area and length, can be determined from the GIS and used in HMS. However, hydrologic parameters, such as time of concentration and storage coefficient, should be based on existing HEC-1 hydrologic models provided by the Galveston District and Harris County Flood Control District (HCFCD).

- A ModClark cell parameter file shall be generated in SHG format for quasidistributed rainfall-runoff transformation. Existing HEC-1 hydrologic models will be reviewed for the development of additional ModClark parameters, such as time of concentration and storage coefficient.

- A background map file shall be generated for visualization of sub-basins and streams within HMS.

Precipitation model

- The storm of October 16–18, 1994, will be studied using NEXRAD radar rainfall data in a precipitation grid called Standard Hydrologic Grid (SHG).

- Radar rainfall connections will be configured to work within HMS; the grid-based NEXRAD radar rainfall will be stored as grids in the Data Storage System (DSS).

Control specifications model

- The period of the analysis will be based on the storm events of October 16–18, 1994, with the proper computational time interval to capture the peak flows.

LITERATURE REVIEW AND GIS SUPPLEMENTAL BACKGROUND INFORMATION

The watershed description and its conveyance system have been distilled from previous project reports, correspondence with Galveston District and HCFCD, hydrologic models, and watershed delineation maps. While the literature review was helpful, field observations conducted on May 16 and October 12, 1998, provided many clarifications about the stream networks and their drainage facilities. The information gathered from field observations has been geographically referenced and incorporated in the GIS spatial database for visualization and documentation. The GIS spatial database supported overlays of aerial photography with streams and gage locations, providing valuable supplemental information.

WATERSHED DESCRIPTION AND CONVEYANCE SYSTEM

The Buffalo Bayou watershed is approximately 341 miles square and lies primarily in Harris and Fort Bend counties in southeast Texas. The Buffalo Bayou watershed consists of two major drainage areas, the Addicks Reservoir and Barker Reservoir watersheds, as shown in figure 2. In moderate to larger storms, however, the Buffalo Bayou watershed also drains portions of the Cypress Creek watershed that overflow south into the Addicks Reservoir watershed. The excess rainfall over the Buffalo Bayou watershed is conveyed slowly over the low relief landscape. When it enters the streams, it often travels through a series of hydraulic structures through the urban area before it can exit through the electronically controlled conduits installed in the Addicks and Barker reservoirs. Primarily for flood control, the Addicks and Barker dams form detention reservoirs that have been used to regulate flow releases downstream into Buffalo Bayou Creek. Downstream of both dams, Buffalo Bayou Creek's nondamaging capacity has been established to be 2,000 cfs. The drainage area downstream of the dams is referred to as Buffalo Bayou Local. Buffalo Bayou Creek then flows through the Houston Ship Channel for about 32 miles and becomes a tributary of the San Jacinto River. From a land-use perspective, the condition of the Buffalo Bayou watershed was undeveloped as of 1977. In contrast, the tremendous growth rate of the Houston area is extending suburban development into these watersheds in recent years.

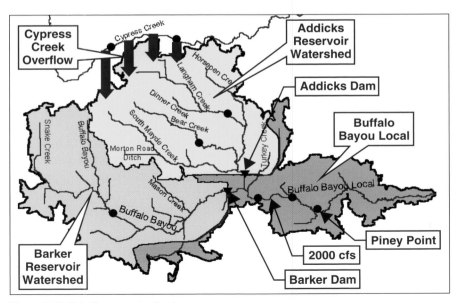

Figure 2. Buffalo Bayou watershed.

Addicks Reservoir watershed

The Addicks Reservoir watershed of about 134.4 square miles lies mostly within Harris County and a small area of about 18 acres lies within Waller County. The watershed slopes about 0.10 percent from northwest to southeast. Ground elevations vary from 163 feet mean sea level (msl) with the 1973 adjustment along the northwestern boundary of the watershed to 71.1 feet msl at Addicks Dam. The mean annual rainfall is 45.75 inches per year. The soil type is classified as Hydrologic Soil Group D, which has high runoff potential and very slow infiltration rates. The runoff generally travels slowly overland due to mild slope, collects in streams, accumulates in the Addicks Reservoir, and discharges through electronically controlled gated conduits into Buffalo Bayou Creek. Table 1 shows streams and drainage areas that have been identified for the study.

TABLE 1. MAJOR STREAMS IN THE ADDICKS RESERVOIR WATERSHED

Major streams	Drainage area (sq. mi.)
South Mayde Creek	31.0
Bear Creek	22.9
Langham Creek	36.8
Dinner Creek	8.2
Horsepen Creek	19.5
Turkey Creek	7.4
Ditch along Morton Road	8.6

Barker Reservoir watershed

The Buffalo Bayou watershed above Barker Dam lies within Harris, Waller, and Fort Bend counties. The watershed is about 124.5 square miles. Natural ground elevations vary from 200 feet msl at the upstream divide to about 73 feet msl at Barker Dam. Natural streamflow gradients in the basin are uniform at about 5 feet per mile sloping in a southerly direction. Rainfall runoff travels slowly overland due to mild slope, collects in streams, accumulates in the Barker Reservoir, and discharges through electronically controlled, gated conduits into Buffalo Bayou Creek.

The Barker Dam was designed and constructed solely for flood control purposes in 1945 by the U.S. Army Corps of Engineers. Both the Addicks Dam and the Barker Dam consist of long earthen embankments with five gated conduits capable of discharging floodwaters into downstream channels. Table 2 shows the streams and drainage areas that have been identified for the study.

TABLE 2. MAJOR STREAMS IN THE BARKER RESERVOIR WATERSHED

Major streams	Drainage area (sq. mi.)
Buffalo Bayou Creek	78.6
Mason Creek	9.4
Snake Creek	36.5

Cypress Creek watershed

The Cypress Creek watershed of about 130 square miles lies to the north, adjacent to the Addicks Reservoir watershed as shown in figure 9. The Cypress Creek channel flows easterly to its confluence with the San Jacinto River in Harris County. Only a few feet higher than the stream bank, the shallow Cypress Creek's floodplain has poorly defined watershed boundaries. Therefore, floodwaters from moderate to large rainfall events overflow southward across the divide at multiple locations into the South Mayde, Bear, Langham, and Horsepen creeks of the Addicks Reservoir watershed. This overflow is not returned to the Cypress Creek basin.

Buffalo Bayou Local

Downstream of the dams, the Buffalo Bayou Local flows are contributed to by a significant drainage area of 82.1 square miles as shown in figure 10.

PROJECT DEVELOPMENT PLANS

The HMS development has been focused on building the GIS spatial database with required thematic layers for modeling the terrain and supporting watershed parameters extraction. Programs in ARC/INFO, ArcView GIS, DSS, and other utilities are identified and used to develop inputs for HMS. These inputs are brought into HMS, which performs the hydrologic modeling and computes flows at locations of interest to the projects. The software packages used in this study are summarized in table 3.

TABLE 3. SUMMARY OF SOFTWARE PACKAGES

HMS	The Hydrologic Modeling System (HMS) was designed as part of the U.S. Army Corps of Engineers Hydrologic Engineering Center's (HEC) "Next Generation (NexGen) Software Development Project." With an improved graphic user interface and advanced technical capabilities for rainfall-runoff processes, HMS will replace the commonly used HEC-1. An HMS model contains three data sets: the basin, precipitation, and control specifications file. The basin file contains hydrologic elements, parameters, and their connectivity to the drainage network. The precipitation file contains meteorological data and methods for averaging the data over the subbasin. The control specifications file identifies the time window of the analysis and computational time interval for the simulation.
	HMS also employs Data Storage System (DSS) for storage and retrieval of time-series and grid-based data and parameters, such as NEXRAD rainfall data and snowmelt and soils moisture accounting parameters.
ARC/INFO	Version 7.2.1 or later from ESRI was used for grid data (DEM) assembly and grid projection to SHG rainfall grid type.
ArcView GIS	Version 3.1 or later from ESRI with ArcView Spatial Analyst extension 1.1 and ArcView 3D Analyst 1.0. ArcView GIS serves as the foundation for most of the GIS terrain analysis, data visualization, and documentation.
CRWR-PrePro	CRWR-PrePro version 3.0 from CRWR is a system of scripts and associated tools that runs within ArcView GIS with ArcView Spatial Analyst. Using the grid-based terrain model built from the DEM, CRWR-PrePro delineates streams and watersheds, determines their interconnectivity, and calculates many physical and hydrologic parameters. CRWR-PrePro prepares an ASCII file in HMS basin format for transferring GIS results into HMS.
GridParm in ARC/INFO	GridParm is implemented as a system of ARC/INFO macros that derive a grid cell-parameter file required by the ModClark quasidistributed method of simulating basin runoff with radar rainfall measurements. The cell-parameter file consists of hydrologic properties of cells defined by the intersection of watershed boundaries with a precipitation-reporting grid in either the National Weather Service's Hydrologic Rainfall Analysis Project (HRAP) grid or the Standard Hydrologic Grid (SHG). Currently, the cell-parameter file contains the following parameters: • Cell identification or coordinates in either the HRAP or SHG • Cell area within the subbasin • Average flow-path length from cell to subbasin outlet The cell-parameter file has been expanded to meet new parameter requirements, such as grid-based curve number and snowmelt parameters. As an example, SCSParm is another ARC/INFO program that works in conjunction with GridParm to extend the cell-parameter file to include curve numbers at the cell level.
ModClark implementation in ArcView GIS	A prototype version of ModClark procedures has been developed in ArcView GIS during this project.
Data Storage System	Data Storage System (DSS) is used for storing, retrieving, and accessing long periods of time-series data, such as streamflow and point precipitation. In this project, a specialized DSS was used to store grid-based radar rainfall. A utility, Ascii2DSS, was used to bring slices of radar rainfall into DSS every five minutes. The DSS stores over 300 radar scenes of the October 16–18, 1994, storm over the entire study area.

The roles of the GIS include terrain analysis, data visualization and exploration, and project documentation. Fulfilling these roles requires that all thematic layers be in the same coordinate system and projection. The selected projection coincides with the SHG projection as described in table 4. Based on the Albers equal area projection, the SHG projection offers many advantages in hydrologic modeling, such as uniform volume of water on a cell basis.

TABLE 4. STANDARD HYDROLOGIC GRID DEFINITION

Units	Meters
Datum	North American Datum, 1983 (NAD83)
1st standard parallel	29 degrees 30 minutes 0 seconds North
2nd standard parallel	45 degrees 30 minutes 0 seconds North
Central meridian	96 degrees 0 minutes 0 seconds West
Latitude of origin	23 degrees 0 minutes 0 seconds North
False easting	0.0
False northing	0.0

Once the GIS data sets are organized in the proper coordinate system and projection, a topographically driven flow network can be modeled with the GIS procedures. More importantly, these GIS procedures can prepare HMS inputs, such as the basin, precipitation, and ModClark files, which are time-consuming, inconsistent, and error-prone if prepared by manual methods.

DATA DEVELOPMENT

The compilation of GIS data sets requires conversion of file formats and coordinate systems, as well as geographical referencing of non-spatial data sets. The files are converted to the industry-standard shapefile format for vector data and ESRI's ARC GRID raster data format, which can be used interchangeably between ArcInfo and ArcView GIS. Triangulated irregular network (TIN) is ESRI's standard format for the ArcView 3D Analyst™ extension. Examples of vector data that require conversions are digital line graphs for stream alignment and State Soil Geographic Data Base (STATSGO) data for hydrologic soil types. Examples of raster data that require conversion are DEM, NEXRAD radar rainfall, and curve number grid. A TIN can also be created to represent the terrain.

It is important to note that these data sets have been reprojected into the SHG projection to ensure that all data sets in vector, raster, and TIN are in proper alignment and refer to the same ground surface. For example, the DEM, originally in the Universal Transverse Mercator (UTM) projection, was reprojected into the SHG projection; nonspatial data sets, such as aerial image and streamflow gage, were geographically referenced to the SHG. Most of the information was acquired online through government Web sites on the Internet as shown in table 5. A few data sources such as NEXRAD rainfall and the SHG curve number grid were obtained from local authorities and academia. The GIS data development for the study area includes all of the Buffalo Bayou watershed, downtown Houston, and most of greater metropolitan Houston.

DATA PREPARATION

The primary GIS data sets as input to CRWR-PrePro consisted of the DEM, DLG, and streamflow gage locations.

Digital elevation model

The DEM was converted from the USGS 7.5-minute quad maps. The 48 tiles of DEM were joined together by ARC/INFO using the ARC GRID extension. The resulting DEM had data gaps along tile edges as shown in figure 3. The data gaps were filled in with interpolated elevation values from their neighboring cells. After verifying the DEM data against the USGS 7.5-minute quad maps, the complete DEM was used to model the terrain as shown in figure 4. From correspondence with HCFCD, the DEM data should be unaffected by the subsidence because it occurred to the east of Houston, which is outside of this project area.

Digital line graph

The DLG represents water features such as streams, ditches, and drainage facilities in more detail than the RF1. Therefore, selective streams and ditches from the DLGs were imposed on the DEM as a way to compensate for the flat landscape. This process is a way to improve the delineation of streams and subbasins.

TABLE 5. GIS SPATIAL DATABASE: DATA AND SOURCES

Data types	Description	Source
Digital elevation model (DEM) data	Originally converted from USGS 7.5-minute quads, the DEM data at 30-by-30-meter resolution is available for the study region from the Texas Natural Resources Information System (TNRIS).	www.tnris.state.tx.us
Spatial rainfall data	Radar rainfall (NEXRAD Stage III) has been compiled for a severe storm during 16–18 October, 1994. At 1-square-kilometer resolution, rainfall intensities are available approximately every six minutes. The rainfall intensity has units of mm/hr.	Mary Lynn Baeck and Jim Smith from the Princeton Environmental Institute
Soil type data	Soil type data under Soil Surveys Geographic Data Base (SSURGO) data is very limited in the study region. Therefore, soil type data under STATSGO data has been selected for the study. The data was collected from 1-by-2-degree topographic quadrangle maps.	United States Department of Agriculture STATSGO CD–ROM
Digital line graph (DLG) data	Digital line graph data is available at 1:100,000 scale from the United States Geological Survey (USGS). This data was derived from 1:100,000-scale, 30-by-60-minute quadrangle USGS maps. The data exists in DLG level 3 format.	edcwww.cr.usgs.gov
Stream coverage	Stream coverage is obtained from the Environmental Protection Agency (EPA) river reach file (RF1). Projected in Albers equal area format, the data has an approximate scale of 1:500,000.	h2O.er.usgs.gov/nsdi/wais/water/rf1.html
Street coverage	Environmental Systems Research Institute, Inc. (ESRI), has compiled a street database for the entire United States. The data is developed from Geographic Data Technology's Dynamap®/1000 street database, version 1, dated 1995. This data can be viewed with ESRI ArcView GIS software.	ESRI Street Database CD–ROMs
Land use/ Land cover	The land-use and land-cover data uses the Anderson Land Use Code (Level II) classification system. At 1:250,000 scale and in Universal Transverse Mercator (UTM) projection, the data was derived from the USGS Houston and Beaumont 1-by-2-degree quadrangle maps. This data was downloaded from the Texas Natural Resources Information System (TNRIS).	nsdi.usgs.gov/nsdi/products/lulc.html
Digital ortho-photo quarter quads (DOQQ)	Digital aerial photos at 30-meter resolution were obtained for the study region from the Texas Natural Resources Information System (TNRIS).	www.tnris.state.tx.us
Streamflow gage data	Streamflow data for the Buffalo Bayou watershed for October 1994 at gages maintained by the USGS.	www.usgs.gov
Curve number grid	Curve number data that is widely accepted by the water resources community in Texas.	CRWR of the University of Texas at Austin
Drainage facilities photographs	Photographs were taken looking upstream and downstream and at faces of accessible drainage structures.	Field observations on May 16 and October 12, 1998

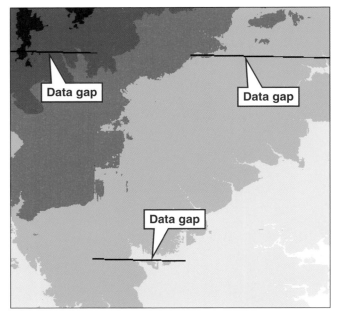

Figure 3. DEM terrain with data gaps.

Figure 4. Complete DEM terrain.

Streamflow data

During the modeled storm, the USGS maintained eight flow gages and two reservoir storage gages in the study area. The stream flow gages are located within Harris County. The USGS classifies these stream gages to the Buffalo–San Jacinto Basin with the Hydrologic

Unit Code of 12040104. For some of these gages, historical daily peak flow values have been retrieved. Other gages, however, contain annual peak flow values, which do not have the details required for this study.

DATA VISUALIZATION AND DOCUMENTATION

The GIS provides good data visualization and documentation support for geographic images, hillshades, and nonspatial data integration. For example, the aerial photography in multiband color DOQQ is used as the background for mapping locations of drainage facilities and stream gages with additional nonspatial data attributes as shown in figure 5. All of the photographs of drainage facilities taken on May 16 and October 12, 1998, have been geographically referenced in the GIS spatial database. Some areas were not photographed due to road closures, limited access roads, and private property restriction.

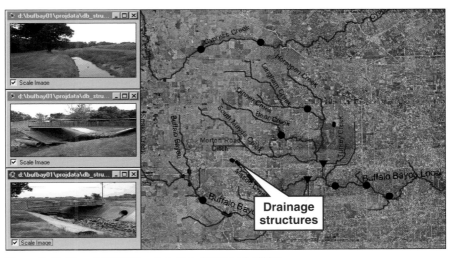

Figure 5. Multiband DOQQ for mapping drainage facilities.

HYDROLOGICALLY CORRECTED AND DEPRESSIONLESS TERRAIN MODEL

Construction of a hydrologically corrected terrain model involves more complexity than combining tiled USGS DEMs into a unifying grid. The DEM assembled from the USGS represented by elevation averages at regular intervals may not accurately represent stream locations and watershed boundaries. For example, stream and watershed delineation does not always coincide with published data sources like the EPA's RF1 and the USGS's watershed in the Hydrologic Unit Code (HUC). A hydrologically corrected terrain model represents accurate flow patterns across landscape, stream alignments, topographic ridges, stream confluences, internal drainage areas, and drainage facilities. Many factors, such as cell resolution, accuracy, topographic relief, and drainage facilities, deserve careful consideration because they often affect the quality of the terrain model. In theory, combining GIS data sets of different resolutions is generally not recommended because of the difficulty in assessing the accuracy and the precision of the resulting data set. In practice, however, combining data sets of various resolutions comes out of necessity due to lack of uniform data and data coverage. For example, DLGs at a scale of 1:100,000 were imposed on the DEM at 30-meter resolution.

In contrast to the effort required for the hydrologically corrected DEM, the depressionless DEM is simply constructed by filling in the sinks or depressions in the assembled DEM. Because of the complexity and effort required for constructing a hydrologically corrected terrain model, a depressionless terrain model often serves as a simpler substitute in the analysis. For study regions with moderate to high topographic relief, the depressionless terrain model may be adequate for the analysis. For low-relief regions, however, the depressionless terrain model often needs additional work to adequately represent the terrain. For example, the Buffalo Bayou region has undergone many iterations, where drainage ditches, particularly the Morton Road Ditch, and streams were imposed onto the DEM to reach greater hydrologic correctness.

GIS APPROACH

The GIS approach toward hydrologic analysis requires a terrain model that is hydrologically corrected. A depressionless terrain model is used in the analysis. The GIS analyzes the depressionless terrain model by applying the eight-point pour model, where water flows across the landscape from cell to cell based on the direction of the greatest elevation gradient. The process of analyzing the landscape characteristics and slopes for stream networks and subbasin boundaries is presented in table 6. The steps in the analysis include imposing the streams on the flat terrain, filling depressions or pits, calculating flow direction and flow accumulation, delineating streams with an accumulation threshold, defining outlets from stream segments, adding user-specified outlet locations, and delineating the subbasin watershed.

TABLE 6. GIS APPROACH TO TERRAIN PROCESSES

Processes	Output

Step 1
Depressionless terrain

Fills the terrain pits by increasing the elevation of the pit cells to the level of the surrounding terrain in order to determine flow directions. The pits are often considered errors in the DEM due to resampling and interpolating grid. Elevation unit is in meters.

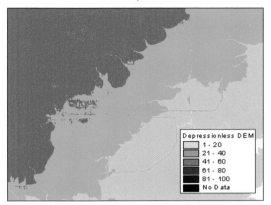

Step 2
Flow direction

Defines the direction of the steepest descent for each terrain cell. Similar to a compass, the eight-point pour algorithm specifies the following eight possible directions:

1 = east, 2= southeast,

4 = south, 8 = southwest,

16 = west, 32 = northwest,

64 = north, 128 = northeast.

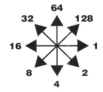

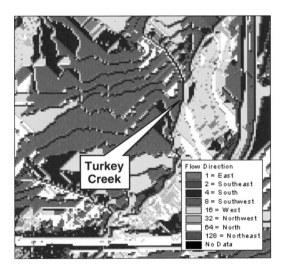

Turkey Creek can be seen flowing south according to the flow direction of 4 (blue cells).

TABLE 6 (CONTINUED)

Step 3
Flow accumulation

Determines the number of upstream cells flowing into a given cell. Upstream drainage area at a given cell can be calculated by multiplying the flow accumulation value with the cell area.

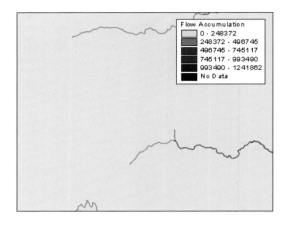

Step 4
Stream delineation

Classifies all cells with flow accumulation greater than the user-defined threshold as cells belonging to a stream network. Typically, cells with high flow accumulation, greater than a user-defined threshold value, are considered part of a stream.

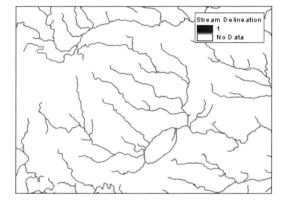

Step 5
Stream link

Stream links or segments are the sections of a stream channel that connect two successive junctions, a junction and an outlet, or a junction and the drainage divide. Two hundred and fifteen stream links make up the stream networks.

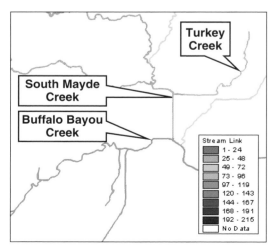

TABLE 6 (CONTINUED)

Step 6
Automatic outlets specification

Outlets are automatically specified as the most downstream cell of the stream segments. Two hundred and fifteen outlets are specified for 215 stream links.

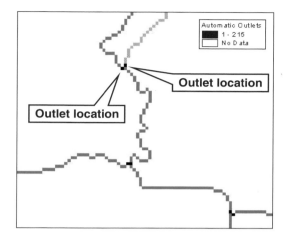

Step 7
User-specified outlet definition

Specifies additional outlets typical of streamflow gages locations and other project control points of interest.

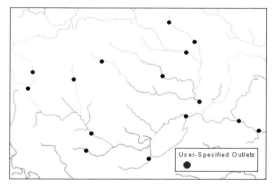

Step 8
Watershed delineation

Delineates the areas draining to outlets.

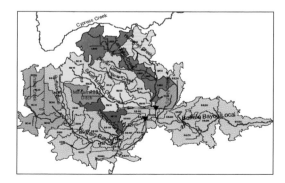

HMS INPUTS

One of the strengths of GIS is its ability to create input files externally that can be imported into HMS. Through spatial analysis, GIS can generate results more efficiently, consistently, and accurately than manual methods. The GIS has produced basin files, background map file, ModClark's cell-parameter file, and grid-based NEXRAD radar rainfall data as shown in figures 6 through 8.

Portions of the HMS basin files in ASCII format showing the drainage patterns with the subbasins, reaches, and junctions and their respective hydrologic parameters are shown in figure 6. The basin model shows the hydrologic methods and their applicable loss rates, unit hydrograph, and routing parameters. Some of these hydrologic parameters were inherited from the GIS and not intended to be used in the final HMS model. These hydrologic parameters will be revised within HMS.

```
Gridded Subbasin: SC10
      Description: Snake Creek
      Canvas X: 9437.440
      Canvas Y: 753347.000
      Label X: 16
      Label Y: 0
      Area: 6.2425
      Downstream: SC-A

      LossRate: SCS
      Percent Impervious Area: 0.0
      Curve Number: 70
      Initial Abstraction: 0.00

      Transform: Modified Clark
      Time of Concentration: 2.42
      Storage Coefficient: 65.3

      Baseflow: None
End:
```

```
Reach: 44
      Description: 225
      Canvas X: 8235.690
      Canvas Y: 748432.140
      From Canvas X: 9006.040
      From Canvas Y: 751267.040
      Label X: -21
      Label Y: 0
      Downstream: SC-B

      Route: Muskingum
      Muskingum K: 0.4732
      Muskingum X: 0.2
      Muskingum Steps: 2
End:
```

```
Junction: BB-G
      Canvas X: 22441.000
      Canvas Y: 738602.440
      Label X: 10
      Label Y: 10
      Downstream: 60
End:
```

Figure 6. Portions of HMS basin files in ASCII format.

The ModClark cell-parameter file showing the x and y coordinates for SHG cells, their average SHG cell (not DEM cell) travel distances, and areas are shown in figure 7.

```
SUBBASIN:   HP10
GRIDCELL:   28    760    4.31344    0.0163324
GRIDCELL:   29    760    3.70745    0.273056
GRIDCELL:   30    760    2.70715    0.552455
                    .
                    .
       Many Lines Omitted
                    .
                    .
GRIDCELL:   33    766    6.40111    0.0675318
GRIDCELL:   31    767    8.31459    0.0024277
GRIDCELL:   32    767    8.04311    0.0761606
END:
```

Figure 7. ModClark cell-parameter file.

The cumulative NEXRAD radar rainfall as shown in figure 8 is in SHG at 1,000-meter resolution. The individual slides of radar rainfall, interpreted for every five minutes, show detailed areal and temporal distribution of rainfall. If the ModClark method is not used, other GIS programs can be used to perform spatial averages and weights based on Thiessen polygons.

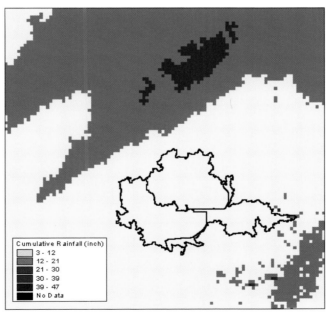

Figure 8. Cumulative NEXRAD radar rainfall.

The verification of GIS results is an important step before using them in HMS as input. The results were verified by graphically checking against the topographic data, street data, and USGS quads; comparing streams and subbasin boundaries with previous HEC-1 model delineation; checking topographic and hydrologic parameters calculations; and comparing drainage area with published information.

Within the GIS spatial database, overlays of pertinent thematic data layers can be viewed simultaneously, facilitating graphical verification. With the subbasin boundaries overlaid on the DEM, the subbasins' shapes and orientations indicate that the flow directions are in agreement with the landscape slope. In addition, the capability to zoom aids in clarifying much drainage detail, such as the flow of Turkey Creek along the toe of the levee and where it joins with South Mayde Creek. After checking GIS results with road maps and USGS quad maps, it becomes evident that the Snake Creek basin is a tributary to the Buffalo Bayou and is incorporated into the HMS model. Graphically checking GIS results through data overlays and comparison with existing maps can be very useful when previous hydrologic studies are not available.

When previous hydrologic studies are available, however, more detailed verification can be performed by comparing GIS topographic and hydrologic parameters with those of the HEC-1 models. In this project, the combination of 30-meter DEM resolution and low-relief terrain can only support broader subbasin delineation, which covers mostly stream confluences and streamflow gage locations. If a DEM is available at finer resolution, more detailed subbasin delineation would be possible; otherwise, attempting to perform more detailed delineation with a 30-meter DEM would likely result in inaccurate subbasin boundaries. Consequently, the subbasins generated from GIS are more generalized and encompass many smaller HEC-1 subbasins; GIS topographic and hydrologic parameters can only be indirectly compared to HEC-1.

Drainage areas and lumped curve number comparison agree for most subbasins. The SHG curve numbers are within the range of the land uses and soil types in the watershed and generally higher for developed areas, as shown in figure 9.

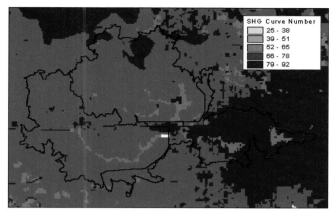

Figure 9. SHG curve number.

Finally, GIS-delineated streams, subbasins, and their associated drainage area were compared with published data sources from the USGS and EPA. For this project, the comparison shows generally good agreement. The GIS-delineated streams consistently align with the USGS's DLG and EPA's RF1 as shown in figure 10. At the Piney Point gage on the Buffalo Bayou Creek, it is important to note that the overall drainage area agrees with published information from the USGS.

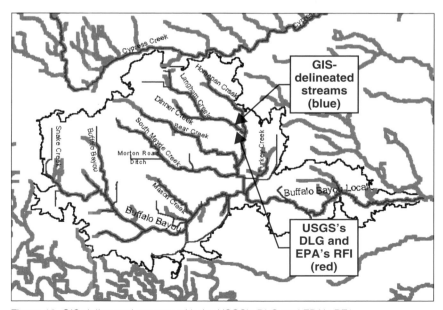

Figure 10. GIS-delineated streams with the USGS's DLG and EPA's RF1.

Using GIS as a preprocessor can accelerate the building of an HMS model. The GIS's ability can extend beyond processing the terrain model to performing spatially intensive analysis for development of grid-based parameters. The results produced by GIS for HMS can be controlled somewhat by focusing on the GIS's description of the landscape characteristics and stream networks. From a modeling standpoint, greater control over the model is often necessary to address difficult situations. HMS is powerful in that it offers full control over the model connectivity, methodology, and parameters. For example, HMS can be used to reconfigure the connectivity and eliminate, add, and revise hydrologic elements and their properties. To model the complex Buffalo Bayou conveyance system, it becomes necessary to take advantage of HMS's flexibility to incorporate the Galveston District's and HCFCD's existing hydrologic and hydraulic models in HEC-1 and Water Surface Profiles (HEC-2) as well as modeling the Addicks and Barker reservoirs. The work going into the three HMS components is discussed below. Some work items are pending because clarifications are required after the Galveston District and the HCFCD have an opportunity to review this report.

The HMS project definition shows the basin, precipitation, and control specification components, which can be selected to create a run as shown in figure 11.

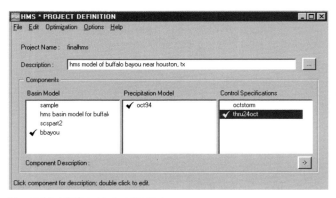

Figure 11. HMS project definition.

BASIN MODEL

The basin model generated by GIS was imported into HMS as shown in figure 12. The basin file contains the schematic, the hydrologic methods, and their parameters. The basin schematic adequately represents drainage patterns with the connectivity of the hydrologic elements (subbasin, reach, junctions, reservoir, diversion, source, and sink). When the connectivity was checked in detail, the presence of short reaches near stream confluence made the model unnecessarily more complicated and required parameter refinement. HMS can simplify the model and save the effort of parameter refinement by eliminating these short reaches and reconnecting the hydrologic element to the nearby junction downstream. However, this simplification was not necessary since the HMS model can be run with GIS-generated parameters.

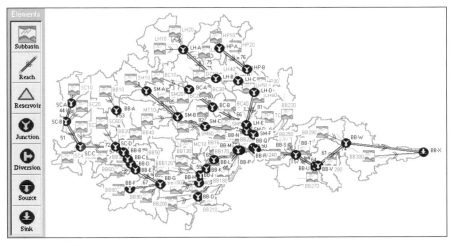

Figure 12. HMS basin model.

The hydrologic methods and parameters have been created by default methods in GIS. The original intention was to revise these hydrologic methods and parameters based on existing hydrologic models in HEC-1 provided by Galveston and HCFCD. The problems that prevented the recycling of these HEC-1 methods and parameters include:

- The subbasin delineation supported by the 30-meter DEM data is less detailed than that of the HEC-1 model. Therefore, migration of HEC-1 parameters requires mapping the HEC-1 delineation over the GIS subbasin delineation. From a visual comparison of the two subbasins' delineation, migration of HEC-1 parameters will be indirect and approximate. Therefore, before additional work is performed, it is important to obtain concurrence on the GIS delineation from Galveston District and HCFCD.

7 The exponential loss rate method in the existing HEC-1 models has been discontinued in HMS. Alternative loss rate methods and parameters that can be derived from GIS spatial data are Natural Resources Conservation Service's basin lumped curve number and SHG grid-based curve numbers.

These obstacles can be overcome through a review process, where decisions can be made by Galveston District and HCFCD on alternative methods and modeling issues.

In the basin model, the ModClark procedures simulate hydrologic processes on a cell basis, which permits more detailed modeling than is possible with lumped parameter methods. The ModClark procedure allows each cell in the grid to have unique values for the parameters, such as travel distance, and a unique value for precipitation depth at each time step. Currently, the ModClark parameters for time of concentration (Tc) and storage coefficient (R) are out of range and need to be estimated from HEC-1 or calibration. The travel distances for each SHG cell in the subbasin are shown in figure 13. The light red cells represent greater travel distances to the subbasin outlet than the dark red cells.

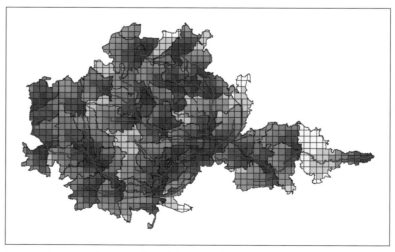

Figure 13. ModClark SHG cells with travel distances.

PRECIPITATION MODEL

The NEXRAD rainfall at stage III indicated that the radar rainfall has undergone automatic and interactive ground truthing with rain gages. The radar rainfall, developed in SHG at 1,000-meter resolution, is stored as a series of grids at five-minute intervals in DSS. The inputs for the precipitation model are shown in figure 14.

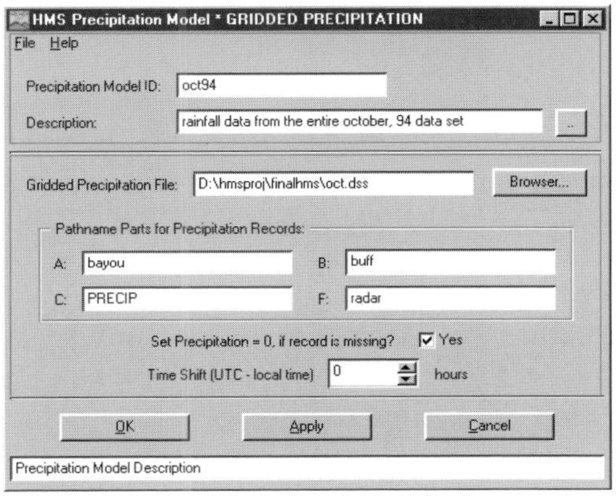

Figure 14. Precipitation model.

CONTROL SPECIFICATIONS MODEL

The control specifications model identifies a time window for October 16–25, 1994. The computational time interval is set at five minutes. The time-related data inputs for the control specifications model are shown in figure 15.

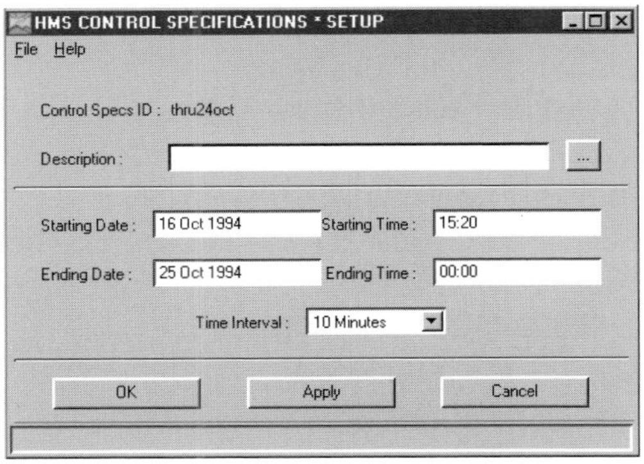

Figure 15. Control specifications model.

HMS RESULTS

An HMS run, shown in figure 16, consists of a basin, precipitation, and control specifications component. Comparison of HMS computed flow hydrograph with the USGS observed streamflow hydrograph at a gage located near West Little York Road on Langham Creek is shown in figure 17. The HMS computed peak flow compares well with the USGS observed flow; however, the HMS model still needs parameter calibration and model refinement for the reservoirs.

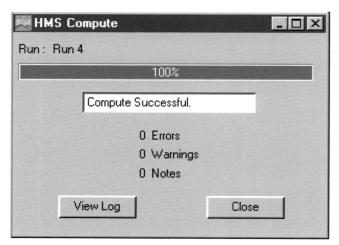

Figure 16. An HMS simulation run.

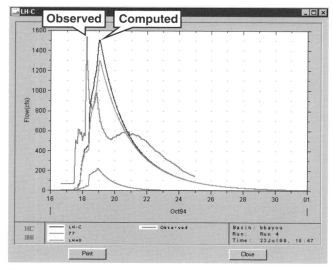

Figure 17. Comparison of computed and observed hydrograph for Langham Creek.

CONCLUSIONS

In the 1970s, the Hydrologic Engineering Center (HEC) participated in developing some of the earliest geographic information system (GIS) applications, such as Hydrologic Parameters (HYPAR) and Spatial Analysis Methodology (SAM), to meet modeling needs in water resources. These early applications were capable of accessing a geographic multivariable grid cell database and developing unit hydrograph parameters. Since then, HEC occasionally participated in the development of GIS applications to keep abreast of the latest technological advances in hydrology and hydraulics. In the 1990s, HEC became aware of the phenomenal growth and advancement in GIS. The capability of obtaining spatial data from the Internet coupled with powerful algorithms in software and hardware made GIS an attractive tool for water resources projects.

The Buffalo Bayou project, funded through the Galveston District of the U.S. Army Corps of Engineers, demonstrates that the development of a quasidistributed hydrologic model in the Hydrologic Modeling System (HMS) is practical with the aid of GIS software and spatial data. The Galveston District plans to use the HMS model for the Buffalo Bayou watershed, located west of Houston, Texas, as the first step in the long-term needs of the Water Control Data System (WCDS). The WCDS is a streamlined collection of programs and hydrologic, hydraulic, reservoir simulation models for real-time monitoring, active warning, and forecasting of floods. The following sections summarize the accomplishments, obstacles, and recommendations for the project.

ACCOMPLISHMENTS

- A GIS spatial database has been built with thematic layers from the U.S. Geological Survey's digital elevation model (DEM) and digital line graph (DLG), U.S. Environmental Protection Agency's river reach file (RF1), and local authorities' loss rate curve number grid and radar rainfall data. With the GIS spatial database serving as a foundation, additional data collected from surveys along major streams was incorporated geographically into the database for documentation, visualization, and modeling.

- An HMS model has been created using GIS to analyze the terrain for determining flow paths and delineating stream networks and subbasin boundaries. In addition, the Modified Clark (ModClark) cell parameter file has been created in Standard Hydrologic Grid (SHG) type for modeling the rainfall-runoff transformation in quasidistributed mode. The NEXt generation RADar (NEXRAD) Stage III precipitation for the storm of October 16–18, 1994, has been configured in the Data Storage System (DSS) to work with HMS for storing and retrieving grid-based time-series rainfall data.

- The ARC/INFO, ArcView GIS, DSS, and other special utilities are time-consuming and require the combined efforts of many people. HEC has worked at integrating existing GIS tools with new programs developed in this project into a comprehensive GIS software package called HEC-GeoHMS. The development of the GeoHMS software is a cooperative effort with Environmental System Research Institute, Inc. (ESRI), through the Cooperative Research and Development Agreement (CRADA), and with the University of Texas at Austin through research contracts.

- With the subbasins delineation being different from that of the existing hydrologic models in Flood Hydrograph Package (HEC-1), HEC has demonstrated how hydrologic parameters from existing HEC-1 models can be brought into HMS for Bear Creek. Review and comments from the Galveston District and the Harris County Flood Control District (HCFCD) regarding the GIS delineated subbasins boundaries and HEC's approach of using of HEC-1 parameters will be important for establishing an acceptable method for the appropriate reuse of existing HEC-1 parameters for the remaining subbasins and routing reaches in the HMS model.

OBSTACLES

- The 30-meter-resolution DEM appears too coarse to support detailed subbasin and stream delineation similar to existing HEC-1 models. In addition, delineation of subbasins and streams requires much iteration because of the low topographic relief. The ability of the GIS program to reliably analyze drainage paths from the relatively low elevation drops between cells in the DEM terrain representation is limited. In many level areas, the stream delineation derived from the terrain does not coincide with published streams from the USGS's DLG and EPA's RF1 in part because many stream widths are less than the resolution of the DEM. To resolve this issue, HEC has imposed selective streams and ditches from the USGS's DLG on the DEM terrain to better reflect the stream locations. This process of imposing streams on the DEM flat terrain, commonly referred to as burning in the streams, can be thought of as laying hot wires (the streams) on a sheet of ice (the DEM flat terrain) and leaving the imprints of the wires on the sheet of ice.

- In addition to the low-relief terrain, which often requires human interpretation of drainage paths, urban drainage facilities and man-made hydraulic structures, such as culverts and underground pipes, dictate flow patterns that cannot be derived from DEM terrain representation. To resolve this issue, HEC has taken advantage of the flexibility in HMS to correct drainage patterns according to human interpretation and local knowledge. HMS is powerful in that it can break faulty subbasin connections and reconnect the subbasins and other hydrologic elements to accurately reflect the watershed drainage patterns.

RECOMMENDATIONS

The GIS spatial database and the HMS model with ModClark quasi-distributed transformation of grid-based NEXRAD radar rainfall serves as an effective foundation for building the remaining models for the WCDS project as shown in figure 18. From the GIS standpoint, a finer resolution DEM should be obtained to accurately represent the low-relief terrain and significantly improve the stream and subbasin delineation. The spatial data layer for the hydraulic structures should catalog other drainage structures for the development of a hydraulic model. From the hydrologic standpoint, the Galveston District should review the approach taken in reusing HEC-1 parameters as demonstrated with Bear Creek. When concurrence on the approach is reached by Galveston District and HCFCD, Galveston District should continue to recycle HEC-1 parameters, or otherwise develop more refined HMS parameters for the remaining subbasins and streams. Finally, Galveston District should assess the value of the Reservoir Simulation System (RSS) for modeling the operations of the Addicks and Barker reservoirs because RSS has been streamlined to be compatible with the WCDS implementation.

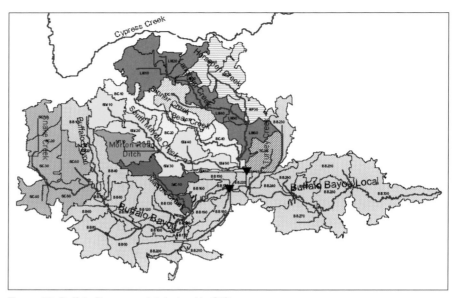

Figure 18. Buffalo Bayou model derived in GIS.

ACKNOWLEDGMENTS

The Buffalo Bayou project has revealed numerous benefits of integrating GIS technology to solve water resources issues. These benefits will lead to improved modeling techniques for studying the watershed and its conveyance system, expedient extraction of watershed parameters, accurate visualization and documentation of the watershed infrastructures and conditions, and effective communication tools. The management people responsible for uncovering these benefits by providing funds, vision, and supervision are Charles Scheffler, chief of the Hydrology and Hydraulic Branch at Galveston District of the U.S. Army Corps of Engineers; Arlen Feldman, chief of the Research Division at HEC; Darryl W. Davis, director of HEC; and David Maidment, director of the Center for Research in Water Resources (CRWR) at the University of Texas at Austin.

The expertise and valuable contributions offered by Seth Ahrens of the Center for Research for Water Resources of the University of Texas at Austin and Harris County Flood Control District are greatly appreciated.

REFERENCES

Addicks Reservoir Watershed Management Study, Phase II Report. 1992. Prepared for Harris County Flood Control District by Bernard Johnson Incorporated in association with Brown & Gay Engineers, Inc., April 1992.

Buffalo Bayou and Tributaries, Texas, Addicks and Barker Reservoirs. Hydrology, U.S. Army Engineer District, Galveston, Corps of Engineers, Galveston, Texas, August 1977.

Hydrologic Engineering Center (HEC). 1998. *Hydrologic Modeling System, User's Manual, Version 1.0.* U.S. Army Corps of Engineers, Davis, California.

Hydrologic Engineering Center (HEC). 1996. *GridParm User's Manual – Procedures for Deriving Grid Cell Parameters for the ModClark Rainfall-Runoff Model.* U.S. Army Corps of Engineers, Davis, California.

ABOUT THE AUTHOR

James H. Doan, P.E. is a licensed civil engineer in the state of California. At the Hydrologic Engineering Center, Mr. Doan is a hydraulic engineer in the Research Division. He works primarily in the applications of hydrologic and hydraulic modeling to flood control projects. He specializes in the integration of GIS with the latest advances in hydrologic engineering and computer sciences. He also provides training courses, lectures, and technology transfer presentations in the area of GIS and hydrology. Prior to joining the Hydrologic Engineering Center, Mr. Doan worked with a private consulting firm in Sacramento performing watershed planning, hydrologic and hydraulic modeling, groundwater modeling, flood protection and control projects, software development and database applications, and design of structures. The products of his work have been used by the U.S. Army Corps of Engineers, Federal Emergency Management Agency, local city and county governments, and private firms to determine flood insurance rates, design and build flood control facilities, and provide water supply alternatives through the conjunctive use approach. Mr. Doan graduated from the University of California, Davis, with a B.S. in civil engineering in 1994.

CONTACT THE AUTHOR

James H. Doan
U.S. Army Corps of Engineers
Hydrologic Engineering Center
609 Second Street
Davis, CA 95616-4687
Telephone: 530-756-1104
Fax: 530-756-8250
james.h.doan@usace.army.mil

Paper 7 **Development of Digital Terrain Representation for Use in River Modeling**

STEPHEN W. LONG

U.S. ARMY CORPS OF ENGINEERS

PHILADELPHIA, PENNSYLVANIA

ABSTRACT

TRIANGULAR IRREGULAR NETWORK (TIN) data structure has the ability to precisely represent linear features (banks, channel bottoms, ridges) and point features (hills, sinks) important in defining channel and floodplain geometry. Both ARC/INFO and ArcView GIS tools have been developed to take advantage of TIN terrain representation to support the HEC-RAS hydraulic model for open channel flow analysis. This paper describes techniques that can be used to develop quality TINs for automated floodplain delineation with HEC-RAS. The focus is on ARC/INFO methodology for underwater terrain modeling based on cross-section data in HEC-2 format (historic hydraulic modeling data).

INTRODUCTION

With the advent of new hydrologic modeling packages, such as HEC-RAS, for performing flood analysis, the representation of a digital surface becomes critical. The major way to represent a three-dimensional surface in a digital format is with a digital terrain model (DTM). The two major DTM formats for representing the surface is the ARC GRID or lattice and the triangulated irregular network (TIN).

Each of these formats can be useful in modeling. The lattice is a grid representation of the surface with a resolution value that represents the cell size and an elevation value associated with each cell. If the resolution is set to a small factor, the surface can be modeled quite accurately. The downside to using lattices is that as the resolution gets smaller the file sizes get larger. TINs are a collection of strategically sampled points and lines, selected to represent key locations on the earth's surface. Points are used to represent hilltops, valleys, and general surface changes. Breaklines are used to control the smoothness and continuity of the surface. These lines can represent such features as ridgelines, river banks, or road features. Of the two DTM formats, the TIN allows for a more precise representation of the surface and can be an important data tool for floodplain geometry.

TIN DEVELOPMENT

TINs can be developed in various ways. The more common method is through photogrammetry. One limitation of existing photogrammetric methods is the inability to obtain ground elevation information below the surface of a river. The portion of a cross section below the water surface can be critical for accurate hydraulic analysis, especially for larger rivers with significant low flow volume. Traditionally, the modeler has two ways to obtain bathymetric information for the modeled cross sections: obtain a hydrographic survey for each cross section in the model, or manually interpolate sampled hydrographic sections to fit the modeled sections. Individual hydrographic surveys for the channel portion of each section in the hydraulic model are generally cost prohibitive. Manually interpolating channel sections to fit the above-water portion of the modeled sections is very labor and time intensive.

The USACE has developed methods and incorporated them into an ARC/INFO program called CHANNEL to automate the generation of bathymetric (or channel) surfaces along a river reach, requiring only a limited number of channel sections as input. These procedures provide the best of both traditional options. Only a limited number of hydrographic surveys can be obtained, thus helping to minimize costs. Also, by providing a complete surface (both above and below the water surface), manual interpolation is not necessary, thus saving considerable time and effort.

INPUT DATA

Before constructing the river, the follow input data is required:

1 A surface model of the potential floodplain surrounding a river channel. This DTM should be in the form of a TIN.

2 An ARC/INFO polygon coverage representing the extent of the area to be studied. This area should extend to the anticipated area of flooding.

3 An ARC/INFO line coverage representing the stream centerline and banks.

4 An ARC/INFO line coverage representing the location of surveyed cross sections.

5 HEC-2 X1 and GR input cards storing stationing and elevations along each surveyed cross section.

PROGRAM METHODOLOGY

This section provides a general description of the methodology implemented by the CHANNEL program. The program is comprised of 13 individual ARC/INFO ARC Macro Language (AML) scripts, each performing a specific function in the creation of the bathymetric surface model.

These procedures use a surface model of the terrain adjacent to the stream channel and surveyed cross-section information stored in HEC-2 X1 and GR cards to interpolate channel bathymetry between the surveyed cross-section locations. The resulting bathymetric data are stored in an ARC/INFO lattice format.

The following procedures are used to create the output surface model:

1 Import HEC-2 Input X1 and GR cards into an INFO™ data table. The SECNO, STCHL, and STCHR variables from the X1 record are written to the INFO file. Each station STA(n) and elevation EL(n) from each GR card is written to the same INFO file.

2 Build a polygon coverage of the area inside the stream channel. A coverage representing the area inside the stream channel is constructed from the stream banks from the input hydrography coverage and from the map panel boundary coverage. This coverage is used to identify those portions of each cross section that lie within the stream channel and to act as TIN breaklines when generating the output bathymetry surface model.

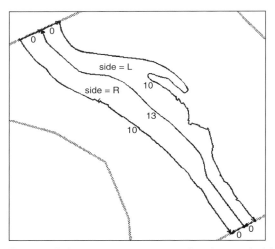

Figure 1. Polygon coverage representing the area inside the stream channel.

3 Build a line coverage containing arcs parallel to the stream banks. These arcs, located inside the stream channel only, are used as TIN breaklines when generating the bathymetry surface model. Each arc vertex is assigned a bottom elevation interpolated from the known bottom elevations at the input cross sections. By using these arcs as TIN breaklines, the bathymetry surface model can be densified so as to minimize the distance between TIN triangle nodes (points with elevations, used to define the surface characteristics).

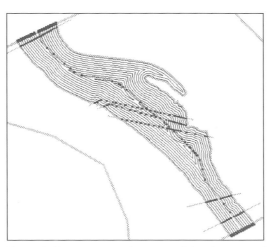

Figure 2. Line coverage of arcs parallel to the stream banks.

4 Identify the portion of each cross section in the stream channel, and associate bottom elevations from the HEC-2 GR cards. The input cross-section coverage is overlaid onto the stream channel polygon coverage created in step 2 to identify those portions that lie within the stream channel. Only those elevations from the HEC-2 GR cards corresponding to stations inside the stream channel are used by the CHANNEL program. All other GR card elevations are disregarded; elevation information on the stream banks is provided by the input digital terrain model.

The GR card stations and elevations stored in the INFO file created in step 1 are associated with the in-channel cross sections via a database relate based on the X1 SECNO value for each cross section. The x,y coordinate for each GR station is established based on its location on the input cross-section arcs, measured from the leftmost point (looking downstream) of each cross section arc. Note: This measurement is based on the leftmost point of the cross section arcs before extracting the portions of the cross section lying within the stream channel. The z value (bottom elevations) for each x,y coordinate is taken directly from the GR records.

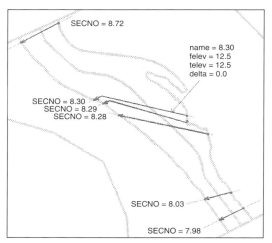

Figure 3. Portion of the cross section within the stream channel.

5 Build end sections, or pseudo cross sections, from those portions of the map panel boundary lying inside the stream channel. Those portions of the map panel boundary lying inside the stream channel are extracted from the stream channel polygon coverage created in step 2. These arcs are assigned bottom elevation values interpolated from the surveyed cross sections located directly upstream and downstream from the map panel boundary. These end sections play an integral role in the generation of the bathymetry surface model as they define the channel bottom elevations at the map panel limits.

6 Interpolate channel bottom elevations at the end sections. Channel bottom elevations along the end sections created in step 5 are interpolated from the GR card elevations at the surveyed cross sections located directly upstream and downstream from each end section. The elevations for the upstream and downstream cross sections, stored in the HEC-2 X1/GR INFO file created in step 1, are associated with the end sections via database relates based on the X1 SECNO value. The interpolated elevation at each x,y coordinate along each end section is determined as a function of the distance of the end section from the upstream/downstream cross sections with known (surveyed) elevations. Note: If the nearest upstream/downstream cross section is located within the current map panel, then the distance to that cross section is measured along the stream centerline arc (from the input hydrography coverage). If the nearest upstream/downstream cross section is located outside the map panel, then the distance to that cross section is based on the SECNO value of that cross section.

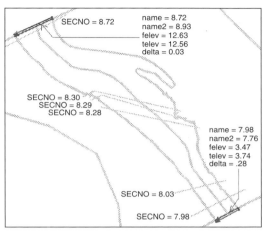

Figure 4. Constructed end sections.

7 Create a TIN breakline file from the stream bank arcs in the input hydrography coverage. This file is in an ARC/INFO generate format and contains x,y coordinates at regular intervals along the stream bank arcs. An elevation (z) value for each x,y coordinate is also written to the breakline file. The elevation at each x,y coordinate is interpolated from the input digital terrain model, which represents stream bank elevations during "normal" flow conditions.

8 Create a TIN breakline file from the stream centerline arc in the input hydrography coverage. This output file is also in an ARC/INFO generate format and contains x,y coordinates at regular intervals along the stream centerline. An elevation (z) value for each x,y coordinate is also written to the breakline file. The elevation at each x,y coordinate is interpolated from a temporary bathymetric surface model. This temporary model is created from the following inputs:

- Bottom elevations along the surveyed cross sections from step 4

- Interpolated bottom elevations at the end sections from steps 5 and 6

- Interpolated stream bank elevations from step 7

The temporary bathymetric surface model is deleted upon establishment of elevations along the stream centerline.

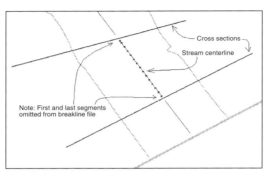

Figure 5. Stream centerline TIN breakline.

9 Create a TIN breakline file from the arcs parallel to the stream banks, as created in step 3. This file is in an ARC/INFO generate format and contains x,y coordinates at each vertex of each arc. An elevation (z) value for each x,y coordinate is also written to the breakline file. The elevations at each x,y coordinate are interpolated from a new temporary bathymetric surface model, created from the following inputs:

- Bottom elevations along the surveyed cross sections from step 4

- Interpolated bottom elevations at the end sections from steps 5 and 6

- Interpolated stream bank elevations from step 7

- Interpolated bottom elevations along the stream centerline from step 8

Similar to step 8, the temporary bathymetric surface model is deleted upon establishment of elevations along the arcs parallel to the stream banks.

10 Create the bathymetric surface model from the following inputs:

- Bottom elevations along the surveyed cross sections from step 4

- Interpolated bottom elevations at the end sections from steps 5 and 6

- Interpolated stream bank elevations from step 7

- Interpolated bottom elevations along the stream centerline from step 8

- Interpolated bottom elevations along the arcs parallel to the stream banks from step 9

The inputs listed above are added either as mass points or hard breaklines to the bathymetry TIN surface model. The surface is created with the ARC/INFO CREATETIN command.

11 Merge the bathymetric surface model with the input surface model of the surrounding floodplain to create the output surface model. The surface model representing bathymetry within the stream channel is merged with the input terrain model, replacing the representation of the water surface in the input model with the representation of channel bathymetry created in step 10. In order to facilitate this operation, both surfaces are converted to an ARC/INFO lattice format using the TINLATTICE command. Surfaces in a TIN format cannot be merged.

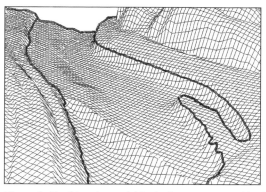

Figure 6. Final terrain model including constructed bathymetry.

When the above procedures are completed, a complete surface model for the study area in a lattice format will exist. This model will be in a lattice format. Complete cross sections can now be obtained at any location along the river reach. By using programs like HEC-GeoRas, these cross sections can be obtained and formatted for HEC-RAS.

CONCLUSIONS

An ARC/INFO methodology for development of underwater terrain representation from HEC-2 cross-section input data has been presented. The underwater data can be merged with the rest of the terrain representation to form a seamless terrain model that can be directly used in a new generation of tools for automated geometry extraction for hydraulic models.

ABOUT THE AUTHOR

Stephen W. Long joined the Floodplain Management Branch of the U.S. Army Corps of Engineers in Philadelphia in 1991. There Mr. Long was responsible for the creation of a geographic information system to assist FEMA in its flood-mapping program. He developed and assisted in the development of several applications, using ARC/INFO software's AML programming language, for automating the production of FEMA's Digital Flood Insurance Rate Maps (DFIRM). He is also involved in other GIS projects within the USACE Philadelphia district, including projects in civil and site engineering, hydrology, and surveying and mapping. In addition, he serves as a member of the Civil Site Field Working Group under the Federal Geographic Data Committee Geospatial Advisory Group. Mr. Long received a bachelor of science degree in electrical engineering in 1986 from Temple University.

CONTACT THE AUTHOR

Stephen W. Long
U.S. Army Corps of Engineers
Wanamaker Building
100 Penn Square East
Philadelphia, PA 19107-3390
Telephone: 215-656-6552
Fax: 215-656-6820
stephen.w.long@usace.army.mil

Paper 8

HEC-GeoRAS: Linking GIS to Hydraulic Analysis Using ARC/INFO and HEC-RAS

CAMERON T. ACKERMAN

THOMAS A. EVANS

GARY W. BRUNNER

U.S. ARMY CORPS OF ENGINEERS—HYDROLOGIC ENGINEERING CENTER

DAVIS, CALIFORNIA

ABSTRACT

ALTHOUGH HYDRAULIC ANALYSIS can benefit from digital terrain models and other GIS data sets, the complexity and unfamiliarity of GIS programs deters hydraulic engineers from using them. In developing HEC-GeoRAS, the Hydrologic Engineering Center has linked ARC/INFO software's data development and display capabilities to HEC-RAS's hydraulic analysis capabilities. GeoRAS allows an engineer with little GIS training to use ARC/INFO to develop geometric data for import in HEC-RAS and view exported water surface profile data. The GeoRAS interface provides specific and logical access to ARC/INFO, allowing the engineer to concentrate on hydraulic principles, rather than GIS mechanics, during model development and analysis.

INTRODUCTION

Geographic information systems (GIS) provide a powerful tool for storing, managing, analyzing, and displaying spatially distributed data. Unfortunately, due to the complexity of GIS, the benefits from using a GIS to assist in water-resource analysis are not available to the inexperienced or infrequent user. Many engineers have had little or no training working with GIS technology, and to perform even simple tasks within a GIS may prove quite burdensome. In the end, the benefits derived by engineers from complex data operations are often dwarfed by the cost of acquiring the knowledge to execute tasks within a GIS.

Hydraulic modeling of river systems may be greatly facilitated with a GIS. Data preparation and model interpretation are time-consuming tasks that can be simplified using a GIS for data processing and display. For the task of river analysis, specific features of a GIS may be made available to the hydraulic engineer through an interface. Providing limited but logical access to a GIS allows the engineer to concentrate on hydraulic concerns, rather than focusing on the complications of applying the GIS.

The development of HEC-GeoRAS has linked ARC/INFO to the U.S. Army Corps of Engineers Hydrologic Engineering Center River Analysis System (HEC-RAS). HEC-GeoRAS provides a graphical user interface (GUI) that enables the hydraulic engineer to create a HEC-RAS import file containing geometric attribute data from an existing digital terrain model (DTM), process water surface profile data exported from HEC-RAS, and perform floodplain mapping.

HEC-RAS

HEC-RAS is a one-dimensional, steady-flow, water surface profiling program (U.S. Army Corps of Engineers, 1998). The HEC-RAS model construct requires definition of the land surface to be modeled and flow data for hydrologic events. The geometric and flow data is used to calculate steady, gradually varied flow water surface profiles from energy loss computations. HEC-RAS is capable of modeling a full network of channels, a dendritic system, or a single river reach.

DATA REQUIREMENTS

HEC-RAS requires the input of geometric data to represent river networks, channel cross-section data, and hydraulic structure data such as bridges and culvert data. At Version 2.0, HEC-RAS introduced the capability to use three-dimensional geometry for the description of river networks and cross sections. Hydrologic events are represented by flow data.

River networks

River networks define the connectivity of the river system. A river system is a collection of reaches, all oriented downstream. A reach is defined in HEC-RAS as starting or ending at junctions—locations where two or more streams join together or split apart. A river may be composed of one or more reaches. Depicted in figure 1 is a general example of a river network with cross-section locations.

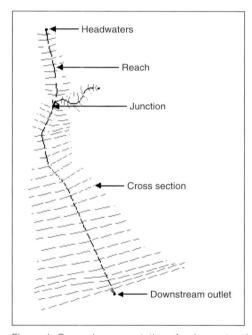

Figure 1. General representation of a river network and floodplain.

Cross-section data

Channel cross-section data is used in HEC-RAS to characterize the flow-carrying capacity of the river and adjacent floodplain. Cross-section data includes station-elevation data, main channel bank stations, downstream reach lengths, roughness coefficients, and contraction and expansion coefficients. Figure 2 provides an example cross section, as represented in HEC-RAS, identifying bank stations and reach lengths.

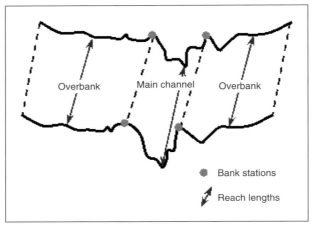

Figure 2. Illustration of channel geometry.

Station-elevation data represents the ground surface at designated locations in a river reach. Cross sections are taken perpendicular to the direction of flow both in the main channel and in the overbank areas. Bank stations separate the portion of the cross section that is the main channel of the river from the adjacent floodplain areas, termed the left and right overbank areas.

Reach lengths are used to define the distance between cross sections and for energy loss calculations in HEC-RAS. Reach lengths are considered for the left overbank, main channel, and right overbank areas, and indicate the path of flow between cross sections.

Roughness coefficients are an indication of the relative channel roughness. Channel roughness is considered for calculating frictional energy loss between cross sections. Typically, channel roughness is indicated by Manning's n-values. Contraction and expansion coefficients are flow dependent and characteristic of abrupt changes in flow direction.

Hydraulic structure data

Hydraulic structure data includes data such as bridges, culverts, and weirs.

Flow data

Hydrologic events are represented by flow data. Flow data includes both the flow magnitude (which may vary on a cross-section-to-cross-section basis) and boundary conditions. Boundary condition requirements will depend on the flow regime of the model.

SIMULATION RESULTS

HEC-RAS is capable of modeling subcritical, supercritical, or mixed-flow regimes. Hydraulic calculations are performed at each cross section to compute water surface elevation, critical depth, energy grade elevation, and velocities.

HEC-RAS IMPORT/EXPORT

Topographic data can be imported into HEC-RAS using a data exchange file format developed by HEC. The ASCII file format is based on key headers to identify geometric information, and may be created or read using any GIS. The GIS data exchange file formats appear in appendix B of the *HEC-RAS User's Manual* (U.S. Army Corps of Engineers, 1998).

HEC-RAS import

HEC-RAS (Version 2.2) supports the import of the basic description of the channel geometry, including the structure of the stream network, as represented by interconnected reaches, as well as the location and description of cross sections. The import file contains river, reach, and station identifiers; cross-section cut lines; cross-section surface lines; cross-section bank stations; and downstream reach lengths for the left overbank, main channel, and right overbank. Roughness coefficients and hydraulic structure data are not written to the import file. Figure 3 is a HEC-RAS schematic of an imported river network.

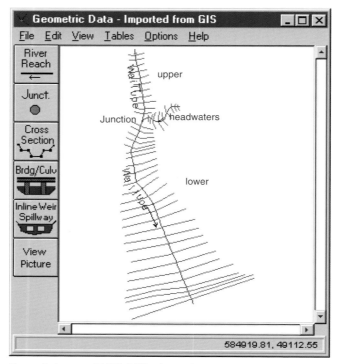

Figure 3. Imported HEC-RAS schematic.

HEC-RAS export

Version 2.2 of HEC-RAS has the capability of writing out simulation results. Cross-section data is written to the export file as cross-section cut lines and cross-section surface lines with river, reach, and station identifiers. Water surface elevations for each profile are attached to each cross section. The export file also contains bounding polygon information for each profile on each reach. Bounding polygon information is used to define the limits of the hydraulic model. A schematic of exported data is shown in figure 4.

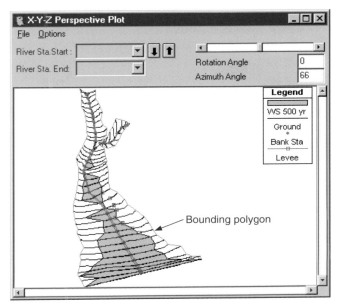

Figure 4. Schematic of exported HEC-RAS results.

GEOMETRIC DATA PREPARATION

A hydraulic model of a river system depends on developing geometric data that accurately depicts the land surface to be modeled. GIS provides the tools for storing and manipulating a three-dimensional representation of the land surface as a DTM, and more specifically to HEC-GeoRAS, as a TIN. Provided a TIN of the channel and adjacent floodplain area, linear data sets (line coverages) can be created based on terrain features and used to develop the required HEC-RAS data components.

Line coverages are data sets composed entirely of lines. In GIS terms, a line is a series of connected points having a beginning and an end. In ARC/INFO, the beginning and end point of each line, or arc, is denoted by a node, while the internal points are termed vertices. Each line also contains attribute data or descriptive information such as length, direction, and connectivity. No two lines in the same data set may intersect.

Geometric data for use in performing river analysis can be developed from a TIN of the land surface and four line coverages. The four line coverages are the Main Channel Invert Coverage, Cross Section Cut Line Coverage, Main Channel Banks Coverage, and Flow Paths Coverage. The purpose for each line coverage in developing geometric data for HEC-RAS is discussed.

MAIN CHANNEL INVERT COVERAGE

The Main Channel Invert Coverage establishes the connectivity of the river system. Each river reach is represented by an arc and is created with its from-node upstream of the to-node. Attributes attached to each arc include the river name, reach name, and stationing of the invert. Each reach on the same river must have a unique name. The river system connectivity and attributes represented by the Main Channel Invert Coverage are depicted in figure 5.

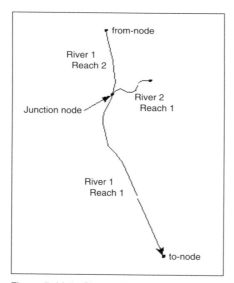

Figure 5. Main Channel Invert Coverage schematic.

CROSS SECTION CUT LINE COVERAGE

The Cross Section Cut Line Coverage is used to define the location of the cross sections and the expanse of the floodplain. Each cut line is constructed from the left bank to the right bank (facing downstream), perpendicular to the path of water flow. Cross-section cut lines are shown in figure 6.

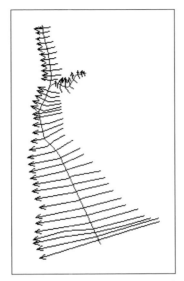

Figure 6. Schematic of cross-section cut lines, drawn from the left overbank to right overbank.

The Cross Section Cut Line Coverage is used to extract cross-section data. The river name, reach name, and stationing for each cross section are determined by intersecting the Main Channel Invert Coverage with the Cross Section Cut Line Coverage.

Station-elevation data is extracted by intersecting the cut lines with the TIN representation of the land surface. Data points having unique values are extracted along the cut line at each break (side of a triangle) of the TIN. Therefore, station-elevation data may be sampled irregularly, depending on the resolution of the TIN. For example, a TIN created from a data set with few mass points in the floodplain, but numerous mass points in close proximity to the main channel, will likely result in a TIN having a dense network near the channel and a less dense network in the floodplain. A visualization of the data extraction process is shown in figure 7.

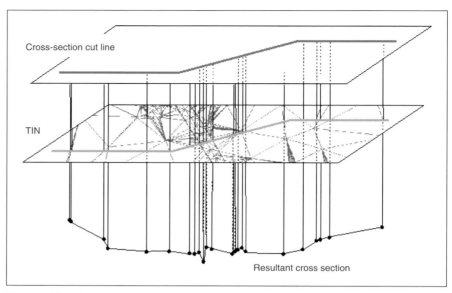

Figure 7. Visualization of station-elevation data extraction from a TIN.

MAIN CHANNEL BANKS COVERAGE

The Main Channel Banks Coverage is used to delineate the portion of cross section that is the main channel from that of the overbank areas. Bank station data is extracted from the intersection of the cross-section cut lines and the bank station lines and expressed as a fraction of the length of the cross-section cut line, as illustrated in figure 8.

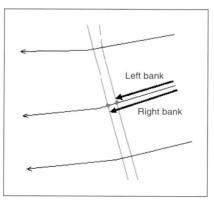

Figure 8. Determination of bank stations from the Main Channel Banks Coverage and cross-section cut lines.

FLOW PATH COVERAGE

The Flow Path Coverage represents the path of water flow in the main channel and in the overbank areas. Flow paths are created in the direction of flow, and indicate the center of mass of flow in the left overbank, main channel, and right overbank. Therefore, the different flow path lines may need to be constructed for different flow events. The flow path lines are intersected with the cross-section cut lines to determine the downstream reach lengths between cross sections, as depicted in figure 9.

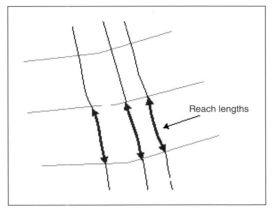

Figure 9. Determination of downstream reach lengths from the Flow Path Coverage and cross-section cut lines.

INUNDATION DATA PROCESSING

Results from HEC-RAS simulations may be written to an export file. The text file includes the locations of the cross-section cut lines along with water surface profile data and a set of polygons that describe the limit of the floodplain modeled. Profile data is written out as water surface elevations attached to each cross-section cut line.

A line coverage of cross-section cut lines is created and attributed with water surface elevations. A TIN of the water surface is then generated from the water surface elevations, using the cut lines from the line coverage as breaklines. Breaklines are used to force the formation of triangle edges during triangulation. Shown in figure 10 is a water surface TIN created using a line coverage for breaklines.

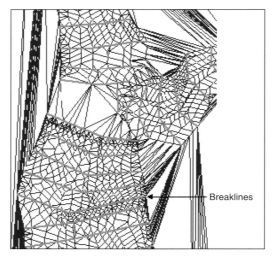

Figure 10. Water surface TIN created using breaklines.

Boundaries for the water surface TIN must then be established to evaluate the extent of inundation. A bounding polygon coverage is created from the boundary data and used to clip the edge of the water surface TIN. The benefits of the bounding polygon are realized between cross sections that are modeled with bridges or levees that are not represented in the DTM. A bounding polygon overlays the water surface TIN in figure 11.

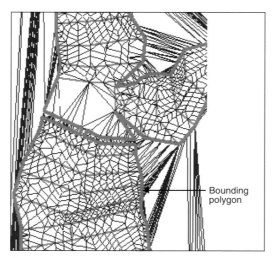

Figure 11. Water surface TIN overlayed by a bounding polygon.

Clipping the water surface TIN with the bounding polygon results in a data set with new boundaries; therefore, a new water surface TIN is created. Figure 12 depicts a water surface TIN created after performing the clipping process.

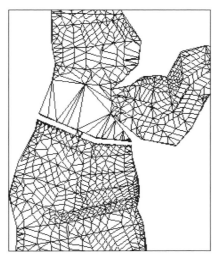

Figure 12. Water surface TIN clipped by a bounding polygon.

Inundation mapping and analysis is then performed by calculations on rasterized (gridded) data created from the water surface TIN and the land surface TIN. The extent of inundation may be viewed as a polygon at the intersection of the water surface and the land surface while an inundation depth may be determined as the difference in elevation between the water surface and the land surface.

**OVERVIEW OF
HEC-GEORAS**

HEC-GeoRAS has been specifically designed to aid engineers with limited GIS experience in the development and display of hydraulic data for HEC-RAS. Knowledge of ARC/INFO, ARCEDIT™, and ARCPLOT™ is advantageous, but not necessary to use GeoRAS. While the GeoRAS user environment has been expressly designed to assist users familiar with HEC-RAS, GeoRAS use is not exclusive to HEC-RAS.

GeoRAS provides the user with environments to manage, create and edit, process, and visualize data sets. Each environment is intended to lead the user through procedures in a clear and logical order as summarized below:

1 Data preparation and management. The user must have a TIN of the river channel and surrounding floodplain.

2 Developing a RAS Import File. RAS Coverages used to extract geometric data from the TIN are created and edited using a data editor. Routines then automate procedures for intersecting the RAS Coverages with the TIN and writing the geometric data to an exchange file.

3 Running HEC-RAS. The hydraulic model must be completed and run within HEC-RAS. A data exchange file of model results is written for import back to the GIS.

4 Viewing water surface profile data. HEC-RAS results are imported to the GIS. Inundation maps may then be created and viewed for further analysis.

DATA PREPARATION AND MANAGEMENT

Data preparation and management is performed using the project manager interface shown in figure 13. The project manager is used to establish and manage projects, define the TIN, and provide access to data preparation procedures. The TIN is the data source from which all geometric data will be extracted to input into HEC-RAS.

Also, the user may have already prepared data sets using a GIS to use in GeoRAS. They may be selected using the project manager.

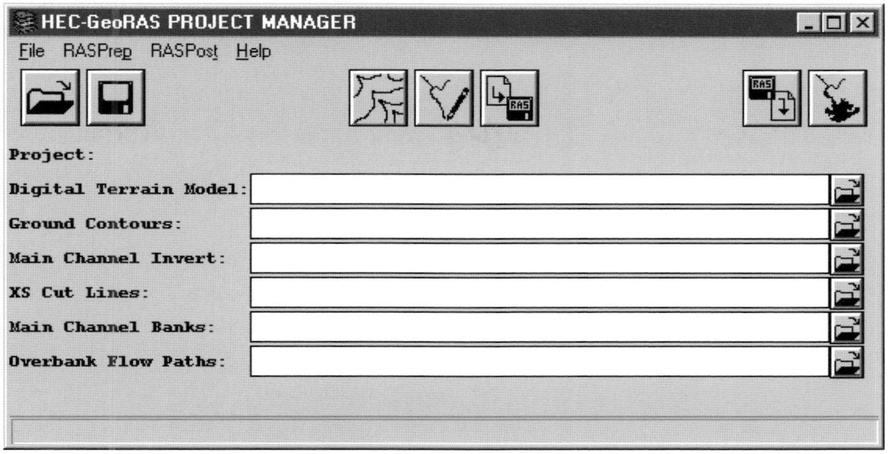

Figure 13. HEC-GeoRAS project manager window.

DEVELOPING A RAS IMPORT FILE

The data sets used to extract geometric data from the TIN are referred to as RAS Coverages. Before accessing the data editor, however, a contour coverage must be created from the TIN to assist the user in developing model data. Contours are displayed instead of the TIN for two primary reasons: the refresh rate for displaying the TIN can be significant, and contours offer an adequate representation of the land surface with relatively few contour lines. The Contour Coverage is used for display purposes only.

RAS Coverage data sets are created and edited using an interactive data editor and display window. The coverage editor is shown in figure 14.

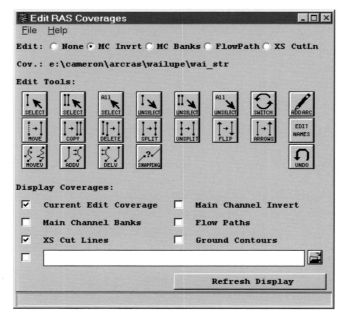

Figure 14. HEC-GeoRAS data editor.

The editor allows the user to create line coverages using a mouse to perform such edits as drawing lines and moving, copying, and deleting lines and vertices, while displaying various line coverages (such as contours) in the background of the display canvas.

The RAS Coverages include data sets that represent the Main Channel Invert, Main Channel Banks, Flow Paths, and Cross Section Cut Lines as shown in figure 15. The purpose of each data set is discussed on the following pages.

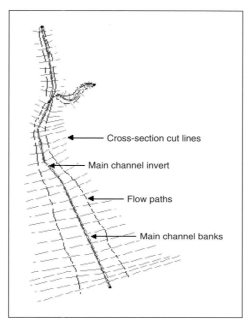

Figure 15. RAS Coverage data used for extracting geometric data.

MAIN CHANNEL INVERT

The Main Channel Invert Coverage is used to define the spatial location (x and y coordinates) of the river and reach network. When intersected with the TIN model, the corresponding elevation will be added to the spatial data. The river network is created with lines oriented downstream and interactively attributed with river and reach names.

MAIN CHANNEL BANKS

The bank station data at each cross section is defined by the Main Channel Banks Coverage. Bank station values are specified by the intersection of the Main Channel Banks data set with the Cross Section Cut Line data set and indicate the bounds of the main channel from the floodplain.

FLOW PATHS

Flow paths define the center of mass of water flow in the main channel and overbank areas. The Flow Path Coverage is used to determine the reach lengths between cross sections in the left overbank, main channel, and right overbank. Flow lines point downstream.

CROSS-SECTION CUT LINES

Cross-section cut lines represent the spatial stationing (x and y coordinates) of cross sections. Cross-section elevation data along each cross-section cut line is extracted from the TIN model. Cross-section cut lines are oriented from the left to right banks, facing downstream.

Creating a geometric data file is automated by HEC-GeoRAS. After each RAS Coverage has been created, routines automatically extract geometric data from the TIN model and data sets and write the data to a data exchange file. Routines validate data sets for completeness and indicate program status to the user.

The import file contains river, reach, and station identifiers; cross-section cut lines, cross-section surface lines; cross-section bank stations; and downstream reach lengths for the left overbank, main channel, and right overbank. Roughness coefficients and hydraulic structure data are not written to the import file.

The user may choose to extract cross-section elevation data at a specified interval or at each triangle edge. The data extraction method selected by the user is based on the TIN model attributes.

RUNNING HEC-RAS

After importing the geometric data extracted from the GIS, completion of the hydraulic data will be necessary. Hydraulic data that is not imported includes roughness coefficients, expansion and contraction coefficients, and hydraulic structure data. Flow data, including boundary conditions, must be specified. After running various simulations in HEC-RAS, water surface profile results are exported to the GIS.

VIEWING WATER SURFACE PROFILE DATA

Water surface profile results from HEC-RAS simulations are imported into the GIS using a data exchange file similar to the geometric data file discussed earlier. During import, polygon coverages and depth grids are created from the water surface profile data.

Inundation data sets are created from water surface elevation data at each cross-section location in concert with a bounding polygon. The bounding polygon limits the area inundated by water due to such features as levees, bridges, and islands, represented in the hydraulic model. Processing of the inundation data sets is described below.

Inundation mapping options are performed using the window interface shown in figure 16 and allow the user to identify the extent and depth of inundation of each water surface. Background coverages may be displayed along with the inundation data to determine areas prone to flooding. Example inundation maps are shown in figures 17 and 18.

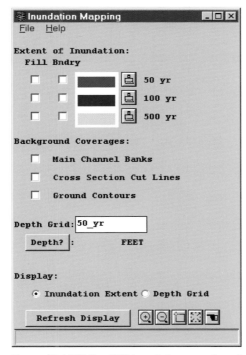

Figure 16. HEC-GeoRAS inundation mapping window.

Figure 17. Inundation polygons for three floods.

Figure 18. Shaded grid of inundation depths.

CONCLUSIONS

HEC-GeoRAS represents a significant advance in linking a GIS with hydraulic modeling. While the integration has been performed specifically for ARC/INFO and HEC-RAS, the GeoRAS software is not exclusive to assisting with hydraulic analysis in HEC-RAS. GeoRAS successfully utilizes the GIS technology to develop and refine geometric data for use in river analysis. Results from hydraulic analysis may be displayed in the GIS to establish floodplain extent and evaluate flood depth.

The user environment exemplifies the integration of a GIS to meet the specific needs of engineers less versed in operating a GIS but for whom the benefits of spatial data development, display, and manipulation in river analysis are invaluable. Providing a simple, logical user environment allows the engineer to view a "real-world" representation of the system under study, and therefore greatly facilitates modeling efforts by assisting the engineer in making informed decisions and identifying errors inherent in model development.

REFERENCES

U.S. Army Corps of Engineers. 1998. *HEC-RAS: River Analysis System User's Manual.* Hydrologic Engineering Center, Davis, California.

ABOUT THE AUTHORS

Cameron T. Ackerman is a hydraulic engineer in the Planning Analysis Division at the Hydrologic Engineering Center, U.S. Army Corps of Engineers. Since 1996, Mr. Ackerman has focused on implementing GIS solutions to problems in surface water hydrology, river hydraulics, and flood damage analysis. He received his master of science degree from the University of California, Davis.

Thomas A. Evans, Ph.D. has led efforts since 1995 to integrate GIS with the hydrologic and hydraulic models developed by the Hydrologic Engineering Center of the U.S. Army Corps of Engineers. Dr. Evans' current work focuses primarily on real-time hydrologic modeling for the Corps's water control activities. He holds degrees in civil and mechanical engineering from the University of Texas and Rice University.

Gary W. Brunner, P.E. is a senior hydraulic engineer in the Training Division at the Hydrologic Engineering Center, U.S. Army Corps of Engineers. Since 1985, Mr. Brunner has provided technical expertise in river hydraulics, unsteady flow and dam break analysis, surface water hydrology, risk and uncertainty analysis, real-time river forecasting, and using GIS information in support of hydrologic studies. He is the team leader for the HEC-RAS (River Analysis System) software development project. He received his B.S. and M.S. from Penn State University and is a registered P.E. in the state of California.

CONTACT THE AUTHORS

Cameron T. Ackerman, Research Hydraulic Engineer
Cameron.T.Ackerman@usace.army.mil

Dr. Thomas A. Evans, Senior Hydrologic Engineer
Thomas.A.Evans@usace.army.mil

Gary W. Brunner, Senior Hydraulic Engineer
Gary.W.Brunner@usace.army.mil

U.S. Army Corps of Engineers
Hydrologic Engineering Center
609 Second Street
Davis, CA 95616-4687
Telephone: 530-756-1104
Fax: 530-756-8250

Paper 9

Floodplain Determination Using ArcView GIS and HEC-RAS

RICHARD A. KRAUS

DODSON & ASSOCIATES, INC.

HOUSTON, TEXAS

ABSTRACT ARCVIEW GIS with ArcView 3D Analyst and a new extension (GIS StreamPro) were used to extract data from a triangulated irregular network (TIN) that represents a watershed being restudied for the National Flood Insurance Program (NFIP). The data consisted of floodplain cross sections and reach lengths, which were located to coincide with field-surveyed cross sections of the channels being restudied. The floodplain data developed in ArcView GIS was imported into HEC-RAS where it was combined with the field-surveyed channel data in order to construct full floodplain cross sections that reflected accurate channel and overbank data for the HEC-RAS model. The HEC-RAS program computed water surface elevations along the channels being studied in this watershed and the results were transferred back to ArcView GIS where the floodplain limits were automatically and accurately mapped.

INTRODUCTION

This paper describes how ArcView GIS with 3D Analyst plus a new extension (GIS StreamPro) were used to facilitate the hydraulic modeling of a channel with several tributaries and the plotting of the resultant floodplain. The hydraulic model used the HEC-RAS (River Analysis System) program developed by the Hydraulic Engineering Center of the U.S. Army Corps of Engineers.

GIS StreamPro is an ArcView GIS extension developed primarily at ESRI under the name AVRAS. Dodson & Associates, Inc., will be distributing GIS StreamPro and providing technical support, training, a comprehensive user manual, and periodic upgrades. GIS StreamPro interprets cross-section data and reach lengths from a terrain TIN (triangulated irregular network), then creates a file with that data that can be read into the HEC-RAS program. The data creates the basic geometry file of a hydraulic model. Bridges, culverts, other physical characteristics of the channels, and overbanks that cannot be extracted from the TIN are manually entered in the HEC-RAS program along with discharges, starting water surface conditions, and other coefficients ("n" values, expansion and contraction coefficients, and so forth). After the HEC-RAS model is completed and the results are acceptable, a GIS export file is created by the HEC-RAS program, which contains centerline and cross-section alignments, water surface elevations at the cross sections, and a bounding polygon defining the maximum extent of the floodplain. GIS StreamPro reads the GIS export file and creates a water surface TIN from the bounding polygon, cross-section alignments, and water surface elevations, then compares the water surface TIN to the terrain TIN and produces a floodplain polygon.

THE STUDY   The watershed analyzed is Mill Creek in Lufkin, Texas. The study consisted of identifying existing flood hazards, and developing both short-term and long-range plans for mitigating those flood hazards. The City of Lufkin and the Texas Water Development Board jointly funded the study. This paper covers the identification of the existing flood hazard areas.

The portion of the Mill Creek watershed being studied covers approximately 3.42 square miles (2,190 acres). The watershed is partially urbanized, especially in the southern portions. The downstream limit of the study is just downstream of Ellen Trout Lake, located near the northern boundary of Lufkin. The study proceeds upstream beyond the lake and splits into the West Branch and East Branch tributaries. Continuing upstream, each branch then splits into East and West forks. The result is a channel system consisting of three junctions and seven reaches: West Branch West Fork and West Branch East Fork combine and flow into the West Branch; East Branch East Fork and East Branch West Fork combine and flow into the East Branch; and the two branches combine and flow into Mill Creek.

A total of 13 roads and railroads cross the channels of Mill Creek and its various branches and forks. The existing channels are for the most part unimproved. The channel sides are steep in many areas, and there is evidence of erosion in some reaches. The banks and bottom of the channel are vegetated with brush and small trees in many areas.

The City of Lufkin is a participant in the National Flood Insurance Program. The Flood Insurance Study (FIS) 100-year floodplain was published in 1982 from a detailed study. It extended about 4,000 feet upstream from Ellen Trout Lake on both East and West branches. This existing-condition study extended the FIS limits approximately 3,000 feet on both branches and added studies of two more tributaries. The figure on the following page illustrates the existing Flood Insurance Rate Map for the portion of Mill Creek being studied.

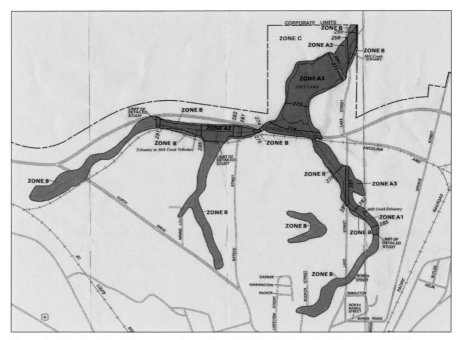

Figure 1. Effective Flood Insurance Rate Map, Mill Creek, City of Lufkin, Texas, 1982.

THE ANALYSIS

The study began with the collection of data, which included previous studies, field reconnaissance, aerial photographs, topographic maps, and so on. Field survey needs were identified, including benchmark identifications and channel cross sections. The channel cross sections were drawn on topographic maps used in the field by the surveyors.

The hydrologic analyses of the Mill Creek watershed used the HEC-1 computer program developed by the Hydrologic Engineering Center of the U.S. Army Corps of Engineers. The watershed was divided into 15 subareas and peak flow rates were computed based on standard design storms, Clark unit hydrograph methods, and modified Puls routing routines.

A comparison of peak flow rates from FIS with the results of the existing-condition study found consistent results. The following table lists the drainage area sizes and 100-year peak flow rates.

TABLE 1. COMPARISON OF 100-YEAR FLOW RATES FROM FIS AND EXISTING-CONDITION STUDY

Location	Flood insurance study		Existing-condition study	
	Drainage area (sq. mi.)	100-year peak flow (cfs)	Drainage area (sq. mi.)	100-year peak flow (cfs)
Mill Creek at corporate limits	3.4	2,830	3.42	2,878
East Branch at Ellen Trout Lake	1.0	1,380	1.10	1,102
West Branch at Ellen Trout Lake	1.9	1,890	1.64	1,700

HYDRAULIC ANALYSIS

The HEC-RAS computer program developed at the U.S. Army Corps of Engineers Hydrologic Engineering Center was used for all hydraulic analyses of Mill Creek and its tributaries. The HEC-RAS program uses Manning's Equation to compute water surface profiles given cross-section data, roughness coefficients, and flow rates. In addition, the program has a number of special capabilities for analyzing culverts and bridges at roadway crossings.

The HEC-RAS data used for all analyses is based on field survey data provided by Everett Griffith Jr. & Associates, Inc. (EGA), and supplemented with data from aerial topographic maps developed by United Aerial Mapping, Inc. Field-surveyed cross sections obtained by EGA typically include the channel plus overbank data for a distance of 200 to 300 feet on either side of the channel. The cross sections were located near the various bridges and culverts and at selected intervals along the channels.

Creating the terrain TIN

The terrain TIN was developed from a combination of field survey data and topographic map data. The survey data was collected on total station survey instruments, which generated a tabular listing of the x, y, and z coordinates for every point. The topographic maps consisted of 2-foot contours supplied in digital format as three-dimensional lines readable in AutoCAD®. A table of x, y, and z coordinates generated from the topographic contours was consistent with the field survey data in terms of horizontal coordinate system and vertical datum. The survey data point listing was then added to the table of points from the topographic maps. In ArcView GIS, an event theme was created from the point table. The terrain TIN was created from a command in 3D Analyst.

Creating two-dimensional lines

The next step in preprocessing a HEC-RAS geometry file is locating various polylines over the terrain TIN where data will be extracted and measurements taken. These lines are two-dimensional and represent the channel centerline, high bank locations, flow path lines, and cross-section locations. Each line or set of lines is a separate polyline-type shapefile theme created for this study.

After the study limits were set along the various tributaries, the centerline theme was created and the centerlines for each channel were drawn from upstream to downstream. The tributary centerlines were also drawn, with the downstream ends snapped to the centerline of the receiving channel; the receiving channel centerlines were segmented into upstream and downstream reaches. The attribute table of the centerline theme was edited by creating columns that identified the "river" and "reach" names (HEC-RAS terminology) for each channel segment. Another column was added that identified each segment as being either an intermediate segment or the furthest downstream segment of a river. The centerline theme is used to determine stationing along the channel and the connectivity of the various reaches.

The high bank lines were drawn according to the terrain TIN for each channel segment. The bank lines are used to determine the locations of the left and right banks on the cross sections.

The flow paths theme consists of three polylines for every channel segment. The lines represent the left overbank flow, main channel flow, and right overbank flow. The center flow path was created by copying the centerline polyline. The left and right overbank flow paths were drawn from upstream to downstream at the assumed location of the centroid of overbank flow, namely one-third the width of the floodplain measured from the channel banks. The attribute table of the flow paths was edited to identify the flow paths as being left, right, or main channel.

The cross-section theme was created and the cross-section alignments were all drawn left to right looking downstream. The locations of the cross sections were drawn approximately where the field survey data for the channels were taken (more about this under Input Data Manipulation) and extended into the overbanks enough to encompass the floodplain. Each cross section traversed one set of flow paths, high bank lines, and one centerline. The cross sections were not allowed to cross other cross sections. The following figure shows the stream layout with the two-dimensional line themes and the direction in which they were drawn.

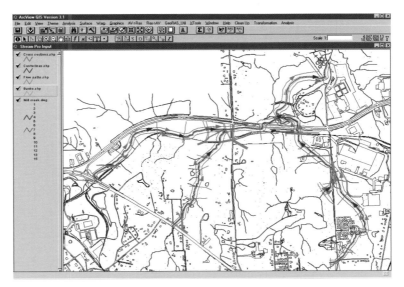

Figure 2. Polyline themes drawn over the basemap.

HEC-RAS GIS import file creation (preprocessing)

After the 2-D lines are drawn and the attribute tables for the centerline and flow paths themes are edited, a number of steps are performed that create the data file used by the HEC-RAS program. These steps are listed under the AV>RAS menu.

The first step identifies the various 2-D themes used by the program. The Theme Set-up dialog box includes entry windows to identify the terrain TIN, 2-D centerline theme, banks theme, flow paths theme, 2-D cross-section theme, and the HEC-RAS GIS import file name. The HEC-RAS GIS import file name must use the .geo extension.

After identifying the various themes and file names, the 3-D channel centerline is created. This is accomplished by first computing the lengths and stationing along the channel centerlines. The command is located under the utility menu because these values can be manually entered if the start and end stations are already known. The centerline topology is then determined, which assigns from-node and to-node attributes for reach connectivity. Finally, the 2-D centerline is draped across the terrain TIN and a new theme is created, *3-D centerline,* which consists of the alignment of the 2-D centerline but with points located everywhere the centerline alignment crosses an angle in the TIN.

The 3-D cross-section theme is created next. The initial steps involved in this process include assigning the river and reach names to each cross section, determining the cross-section stationing, measuring the locations of the left- and right-bank stations along each cross section, and finally computing the flow lengths between each cross section. The last step correlates the alignment of the cross sections with the terrain TIN by extracting the elevations from the terrain TIN and creating a 3-D cross-section theme.

The final preprocessing tasks create the HEC-RAS GIS import file, which contains the header, the centerline coordinate listing, and the cross-section data. These three parts can be performed separately, under the pull-down menu, or automatically by selecting a single tool bar button. The HEC-RAS GIS import file is an ASCII text file.

Input data manipulation

In this study, the cross-section data points collected from the field survey efforts were added to the data point listing of the topographic contours before the terrain TIN was created. This allowed the cross-section alignments to be drawn close to the field survey points. However, it is important in hydraulic modeling that the actual field survey points be reflected in the cross-section data points so that verification of actual field-measured elevations can be made. To do this, the HEC-RAS GIS import file was edited in a text editor by selectively replacing the points taken from the terrain TIN at the channel with actual surveyed points. The result was cross-section point listings that contained data from the terrain TIN for the overbanks and data from the field survey for the channel. The reach lengths and stationing remained unchanged, but the bank locations had to be manually adjusted.

There are other ways of achieving the same results. The cross-section lines can be snapped to the survey points in AutoCAD and copied to the 2-D cross-section theme. With this procedure, manipulation of the HEC-RAS GIS import file would be eliminated.

HEC-RAS modeling

The HEC-RAS GIS import file is read into the geometry data editor. The view displays the centerline and cross sections with the proper alignment and orientation. The river and reach names appear, along with the cross-section stations and flow direction arrows.

After the cross sections were created, 13 culverts and bridges along Mill Creek and its various branches were inserted into the HEC-RAS models. The Manning roughness coefficients for channels and flood-plains were also inserted into the models to complete the geometry models of Mill Creek and its tributaries. Flow rates from the hydro-logic analysis were used in the steady flow files, and normal depth starting conditions were selected, which required the input of the average channel slope. In this study of Mill Creek and its tributaries, there are seven reaches and three junctions. The GIS import file auto-matically establishes the connectivity of these reaches to the various junctions. As a result, the downstream and upstream starting hydrau-lic conditions for the different reaches are automatically set up in the steady flow editor.

The following table compares the effective FIS and the existing-condition 100-year flood levels for Mill Creek and its tributaries. Existing-condition flood levels are similar to or somewhat greater than FIS values at most locations.

TABLE 2. COMPARISON OF 100-YEAR FLOOD LEVELS FROM FIS AND EXISTING-CONDITION STUDY

Location	FIS (feet)	Existing condition (feet)
Mill Creek (called "Mill Creek Tributary" in FIS)		
Upstream of Lake Street	258.40	258.31
Upstream of Ellen Trout Lake Spillway	270.80	270.91
Upstream of Loop 287	273.80	274.28
Mill Creek East Branch (called "Mill Creek Tributary" in FIS)		
Upstream of railroad	274.30	274.90
Upstream of Lake Street	281.00	280.68
Mill Creek West Branch (called "Tributary to Mill Creek Tributary" in FIS)		
Upstream of Loop 287	273.80	274.27
Upstream of Sayers Street	281.60	278.94

HEC-RAS output

After a satisfactory HEC-RAS model is created and the output deemed acceptable, the GIS export function in HEC-RAS is applied. The commands are located under the main window file menu. The options available include naming the export file, selecting the profiles to be exported, deciding whether or not to include interpolated cross sections, and deciding whether to select the entire cross section or just the channel.

The actual export file is an ASCII text file containing a header that includes file names, number of profiles, reaches, cross sections, and so forth. The next section of the output file is the stream network coordinate listing, which includes the coordinates of the nodes for each reach (upstream and downstream end points) and the 2-D cut line for each reach. The cross sections follow the centerline information and contain identification data, water surface elevations, and cut lines. The last part of the output file is the coordinates of the bounding polygon. This polygon defined the maximum possible extent of the floodplain and is defined according to the limits of the cross sections. The coordinates of the polygon correspond to the end points of the cross sections. For floodway profiles, the bounding polygon would correspond to the encroachment stations that define the floodway limits.

FLOODPLAIN MAPPING (POSTPROCESSING)

With the HEC-RAS GIS export file created, the postprocessing aspects of GIS StreamPro were used to map the floodplain over the terrain TIN. In ArcView GIS under the RAS>AV menu, the first command is a Theme Setup dialog box that specifies the HEC-RAS GIS export file, the terrain TIN, the output directory name where the floodplain themes are stored, and a rasterization cell size. The rasterization cell size corresponds to the map units and defines the resolution of the floodplain boundaries.

Postprocessing is performed in either three separate steps or automatically through a tool bar button. The steps involved in the postprocessing start with the generation of the bounding polygon and cross-section alignments. These are read directly from the HEC-RAS GIS export file and new themes are created for each. A water surface TIN is created next, with the extent defined by the bounding polygon. The water surface elevations at each cross section are applied along the cross-section alignments and the TIN is created by treating the cross sections as breaklines in the elevations.

The final step meshed the water surface TIN with the terrain TIN to produce the floodplain. This command converts the water surface TIN and the terrain TIN into grids according to the rasterization cell size specified in the Theme Setup dialog box. The two grids are then compared cell by cell and the cells where the water surface elevations are greater than the terrain elevations are considered to be within the floodplain. After the comparison is complete, the cells considered to be within the floodplain are converted into a polygon shapefile.

The resulting existing-condition floodplain polygon was then placed over the existing Flood Insurance Rate Map (FIRM) to compare the results. The FIRM was scanned and brought into ArcView GIS as an image, then converted to a grid. By using the warp commands, the FIRM grid was manipulated to correspond to the coordinates of the floodplain polygon. The following figure illustrates the comparison.

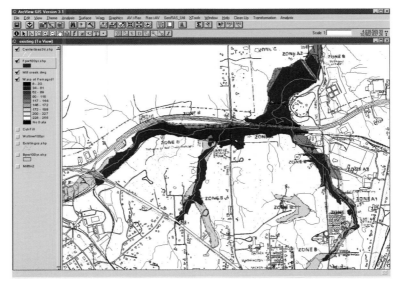

Figure 3. Comparison of existing-condition 100-year floodplain to effective FIRM.

The comparison shows differences between the two floodplain plots. Some reaches have wider floodplains, some narrower.

CONCLUSIONS

The use of GIS StreamPro in an actual study increased efficiency and accuracy. The time required to manually code the cross-section points into the HEC-RAS model was reduced and human errors due to typographical mistakes were eliminated. Additionally, during the fully developed conditions and interim flood control conditions included in the study, time and effort were reduced because the terrain TIN had already been created during the existing-condition analysis.

Not only were human errors eliminated during the input of data, but also during the floodplain plotting. Prior to GIS StreamPro, floodplain plots between cross sections were subject to interpretation of the contours. With GIS StreamPro, the floodplain is plotted continuously according to the terrain TIN and no interpretation between cross sections is needed.

ACKNOWLEDGMENTS

The program development and contributions from the Hydrologic Engineering Center and ESRI that led to GIS StreamPro are greatly appreciated. Without the foresight and commitment of the HEC to create new and better ways of utilizing data sources in the field of water resources, GIS StreamPro would not be possible.

REFERENCES

Barrett, Duane, Dodson & Associates, Inc. 1998. Identification of Existing Flood Hazards and Development of Interim and Future Drainage Improvement Plans for the Mill Creek Watershed. City of Lufkin, December 30, 1998.

Everett Griffith Jr. & Associates, Inc. Field Survey Data for Mill Creek and Tributaries.

United Aerial Mapping, Inc. Aerial Topographic Maps.

Federal Emergency Management Agency. 1982. Flood Insurance Study for the City of Lufkin, Texas.

Brunner, Gary W. 1997. *HEC-RAS User's Manual*.

Djokic, Dean. 1998. AVRAS Extension. Environmental Systems Research Institute, Inc., Redlands, California.

ABOUT THE AUTHOR

Richard A. Kraus, P.E. is a senior hydrologist with Dodson & Associates, Inc., in Houston, Texas. Mr. Kraus has more than 18 years of experience in the management, engineering, and planning of civil projects including flood control studies, development impact and mitigation analyses, watershedwide feasibility, and master plans. He has taught seminars for the American Society of Civil Engineers on such topics as the HEC-2, HEC-RAS, and HEC-HMS/HEC-1 software programs, and the National Flood Insurance Program. He is the product manager of, and has conducted training classes for, GIS StreamPro, an ArcView GIS extension linking ArcView GIS to HEC-RAS.

CONTACT THE AUTHOR

Richard A. Kraus
Dodson & Associates, Inc.
5629 FM 1960 West, Suite 314
Houston, TX 77069-4216
Telephone: 281-440-3787
Fax: 281-440-4742
dkraus@dodson-hydro.com
www.dodson-hydro.com

Paper 10

The Accuracy and Efficiency of GIS-Based Floodplain Determinations

ROY D. DODSON AND XIAOJIAN LI

DODSON & ASSOCIATES, INC.

HOUSTON, TEXAS

ABSTRACT

NEW SOFTWARE AND HARDWARE PLATFORMS, along with new data collection technologies, are making it feasible to prepare digital terrain models (DTMs) for many areas, stored as triangulated irregular networks (TINs). These TINs can be used to compute floodplain elevations and map floodplain boundaries, using new automated floodplain-mapping software.

This paper explores the differences between a floodplain study performed using automated floodplain-mapping software and the same study performed using currently accepted methods. Two questions are answered in this paper:

- Does the use of automated floodplain-mapping software affect the efficiency of a floodplain study?

- Does the use of automated floodplain-mapping software affect the accuracy of a floodplain study?

To answer the first question, a stream channel was studied using traditional methods. The same stream channel was then studied using automated floodplain-mapping software. A careful record was maintained of the tasks performed and the time required for each task. The results indicated that automated floodplain-mapping software provided significant improvements in efficiency for many of the tasks involved in floodplain computations and mapping. Even more dramatic improvements should be expected when existing data have to be revised or corrected.

To answer the second question, the results of the traditional analysis were compared with the results of the automated method. Individual sources of error were isolated and analyzed for each method. The conclusion was that the elimination of almost all manual data entry through the use of automated floodplain-mapping software should result in significantly fewer human errors in the hydraulic analysis and floodplain-mapping procedures. Therefore, the floodplain profiles and floodplain boundaries produced using the automated procedures should be more accurate under most normal conditions, provided that the TIN model available for use in the automated computations is derived from the same topographic data source used for the manual data entry.

GIS CAPABILITIES OF HEC-RAS

In 1995, the U.S. Army Corps of Engineers Hydrologic Engineering Center (HEC) released the first version of HEC-RAS (River Analysis System). HEC-RAS is a next-generation program, implemented under the Microsoft Windows® operating system, and using modern graphical user interface conventions. HEC-RAS is the successor to the HEC-2 computer program, which was the most widely used method of computing water surface profiles, floodplain boundaries, and other information for stream channels.

HEC-RAS offers many new features and conveniences, but one of the most interesting is a GIS data interface. Through this interface, HEC-RAS can receive input data from an external program (such as a GIS). The input data that may be received in this way include the following:

- The topology of an entire stream system (including the main stream and all of the tributary streams), showing the relative location of each stream and the points of confluence. In HEC-RAS, this is called the stream system schematic.

- The cross-sectional shape and size of each stream channel, using three-dimensional cross-section lines that cut each stream channel and the adjacent floodplain areas on each side of the stream channel. The only limit on the number of these cross sections is the capacity of the hardware platform used for the analysis.

- The distances that the flow must travel in the main channel, the floodplain area left of the channel, and the floodplain area right of the channel, from each cross section to the next one downstream. These are called the channel reach length and the overbank or floodplain reach lengths.

- The locations (stations) along each cross section where the stream channel meets the left floodplain and the right floodplain, respectively. These are called the left channel bank and the right channel bank.

These data values represent the bulk of the input data required for a hydraulic analysis. For stream systems without bridges or other special conditions, only the Manning roughness coefficients, the expansion and contraction loss coefficients, the discharge rates, and the hydraulic boundary conditions would be required in order to compute a complete water surface profile. If road crossings are present, they must be added using HEC-RAS.

After the hydraulic analysis has been performed by HEC-RAS, and the computed water surface profiles for the stream system are available, HEC-RAS can export these results back to the external program (GIS).

CAPABILITIES OF GIS STREAMPRO

GIS StreamPro is a software program distributed by Dodson & Associates, Inc., and designed to provide an interface between HEC-RAS and ArcView GIS (with the ArcView 3D Analyst extension). GIS StreamPro supports all of the features of the HEC-RAS GIS interface introduced above, including the following:

- Automated support for creating the topology of a stream network, including stream centerlines and confluences

- Automated preparation and export of three-dimensional cross sections at user-specified locations and alignments

- Automated preparation and export of reach length and bank station data values

- Automated import of HEC-RAS computational results

- Automated delineation of floodplain boundaries, at a level of precision that may be defined by the user

PURPOSE OF THIS ANALYSIS

With the high degree of automation achieved by GIS StreamPro, it would appear logical to assume that a floodplain study might be performed very efficiently using this new method. However, GIS StreamPro requires that a triangulated irregular network (TIN) be available for all of the floodplain areas to be considered in the analysis. The time and effort required to obtain such a TIN, or to assemble one from available data sources, could outweigh the efficiencies of the other processes.

In order to determine if GIS StreamPro provides increased efficiency, a typical floodplain analysis was performed using GIS StreamPro and using traditional methods. A careful record was kept of the work items and staff time required for each method.

It should be noted that there are various kinds of floodplain analyses, and the goals, extent of details desired, and types of topographic data used may vary considerably, as may the types and characteristics of stream channels and floodplains. The floodplain analysis considered in this paper is one of the most common and simple: perform hydraulic modeling using HEC-RAS and produce a floodplain map for a previously unstudied natural stream. The location is a small city in east Texas, considered to be topographically similar to many areas across the United States.

In a floodplain analysis, efficiency means primarily minimizing the expense in acquiring data and the time required for completing the analysis. A complete analysis of efficiency should consider not only the effort and time required to perform a task once, but also the likelihood that some or all tasks may need to be performed more than once. In completing a floodplain analysis, for example, it is often necessary to repeat parts of the process with some revised input to achieve the desired results, such as inserting additional cross sections.

For this analysis, accuracy is defined as the ability to produce a known or standard result, and the ability to reproduce a previous result. Ideally, a comparison of accuracy should involve using each of the two methods to produce a known or standard result. However, this is quite difficult to do in the case of a computed water surface profile and floodplain boundary, for the following reasons:

- For natural channels (channels in which the shape, slope, and alignment of the channel change continually along its length), it is very difficult to obtain a complete historical record of the water surface profile and floodplain boundaries with enough detail to allow us to distinguish between two methods that do not differ greatly in their results. (Please note that we are primarily interested in the analysis of natural channels with floodplains, because the analysis of prismatic channels—typically man-made, with regular shapes, slopes, and alignments—with no floodplain is relatively simple in comparison.)

- Even if a complete historical record of water surfaces were available for a particular stream channel segment, we would still have to know the actual discharge rate corresponding to the measured water surface profile.

We were unable to locate a stream channel for which these data were available for a historical storm. Therefore, we were unable to perform a direct comparison of accuracy. Instead, we have performed an analysis of the potential sources of error within each of the two methods.

FLOODPLAIN ANALYSIS WORK ITEMS

A floodplain analysis involves data collection and preparation, model creation and execution, and floodplain mapping.

The following sections describe each of these categories and compare the work required using traditional methods and using GIS StreamPro.

DATA COLLECTION AND PREPARATION

A floodplain analysis requires the following:

- Topographic data, including stream channel cross sections and reach lengths

- Hydrologic data, including discharge rates for the storms of interest

- Hydraulic data, including loss coefficients and hydraulic boundary conditions

This analysis is concerned only with the collection and preparation of topographic data. The following have been the traditional sources of topographic data:

- Field-surveyed cross sections obtained using conventional field survey instruments. Elevations are often accurate to with ±0.02 or ±0.03 feet, assuming that a professional survey crew is able to work from a good benchmark using good-quality equipment. However, field survey data is relatively expensive, and budget may not always be available for cross sections that are long enough to extend across the entire width of the floodplain and close enough together to accurately define the channel shape, slope, and alignment. Field-surveyed cross sections may be transmitted as reduced field notes or as digital data files.

- Spot elevations from aerial photogrammetry. Aerial photogrammetry is often used for larger project areas. The accuracy of data obtained in this way is determined by the specifications of the aerial photogrammetry contract. Most such contracts reference an accepted map accuracy standard such as the American Society for Photogrammetry and Remote Sensing (ASPRS 1990). The ASPRS standard requires that spot elevations be accurate to within one-sixth of the contour interval. This would provide accuracy of ±0.33 feet for maps with a 2-foot contour interval or ±0.83 feet for maps with a 5-foot contour interval. Spot elevations may be available in digital form or on drawings.

- Interpolation from contour lines. If no better source of data is available, a contour map may be used to determine cross-section coordinates. These contour maps may have been generated by aerial photogrammetry, but no spot elevations may be available in the area of a cross section. The USGS 7.5-minute quadrangle maps provide coverage of practically the entire United States. The accuracy of the map determines the accuracy of the data obtained by this procedure. Maps produced by aerial photogrammetry should meet the accuracy standards specified in the contract. The USGS has accuracy statistics available for its mapping products. This procedure almost always involves making measurements from a paper or Mylar® copy of the map.

Most floodplain studies involve at least some field survey work. Experience has proven that it is difficult to obtain accurate results without at least obtaining field survey data for the channel portion of each cross section. Often, to save money and time, the remainder of each cross section is obtained from aerial spot elevations, contour maps, or both.

New technologies developed over the past several years now offer new alternative sources of topographic data, including the following:

- USGS DEM files. The USGS has undertaken an ambitious program of generating digital elevation model (DEM) files that correspond to each of the 7.5-minute quadrangle maps in the United States. This has resulted in digital topographic data for most areas. The data files include elevation values, rounded to the nearest meter (or to the nearest foot in areas with very little slope), for each point on a 30-meter grid. This means that there are elevation values every 30 meters (about 100 feet) in both east–west and north–south directions. Today's GIS software, including ArcView GIS (with ArcView Spatial Analyst and ArcView 3D Analyst) can easily read USGS DEM files and produce contour maps and other standard topographic mapping products.

- TIN models: A triangulated irregular network (TIN) model is a representation of the earth's surface as a set of fully connected triangles. TINs are more powerful than DEMs in dealing with complex topographic features. There is some controversy among GIS professionals as to whether TINs are more efficient than DEMs at certain operations. For now, however, it is useful to compare the two as follows: TINs are more accurate whereas DEMs are faster and more efficient. GIS StreamPro uses TINs, which may be assembled from many different sources of topographic input, including DEM files, spot elevations from aerial photogrammetry or field survey, or existing contour maps (which must of course be digitized). The accuracy of a TIN can be equal to the accuracy of the data used to create the TIN, although some TINs may be simplified in order to make them faster and more efficient. This simplification may involve leaving out certain data points or using one new point as a substitute for many old points. These kinds of simplifications reduce the accuracy of the TIN.

The recent emergence of new data collection technologies such as LIDAR (light-imaging radar) is making the use of TIN models much more cost-effective. LIDAR typically provides very dense topographic data (coordinates are often available no more than 10 feet apart). These coordinates can be assembled into a very accurate TIN in many cases, although the size of the resulting data file can be a problem.

In addition to the cross-section data, the importance of the reach lengths between each adjacent cross section must not be overlooked. Channel reach lengths are usually measured from CAD drawings. Measurement of overbank reach lengths, however, requires engineering judgment in deciding the paths of overbank flows. This process is most often performed manually by drawing the overbank flow paths on the basemap and measuring the reach lengths by hand or in a CAD program after digitizing the flow paths.

The biggest item in using GIS StreamPro for data collection and processing is the creation of the TIN. However, ArcView 3D Analyst makes this a relatively simple procedure.

Once the TIN is created, cross sections, flow paths, centerline, and bank lines are required to be drawn in GIS StreamPro. Though this process can be tedious, the flexibility of GIS StreamPro allows the user to use existing survey lines and points (from a project layout drawing prepared using a CAD program, for example). This process also enables the user to draw cross-section lines precisely over surveyed points. This ensures that the cross sections interpolated by GIS StreamPro follow the alignment of field-surveyed cross sections, if these have been included in the TIN model.

Once all the lines are defined in GIS StreamPro, it only takes a few mouse clicks to process all the data and export the results to HEC-RAS.

MODEL CREATION AND EXECUTION

After completing all the cross sections and reach lengths, other data must be completed, including contraction/expansion coefficients, Manning's "n" values, and ineffective flow areas. Hydraulic features, such as bridges, culverts, in-line weirs, and levees must also be coded into the model. This step is essentially the same, regardless of whether the data were entered using the traditional method of manual data entry, or whether GIS StreamPro was used to automate the data entry.

After creating the model, a run is executed and the results of the run are evaluated. Adjustments are made as necessary to produce reasonable and accurate results. The final product of this task is the water surface profile to be used in mapping the floodplain.

FLOODPLAIN MAPPING

The goal of this procedure is to produce the floodplain map from the water surface profile calculated by HEC-RAS. It involves transferring the water surface profile to a basemap to show the extent of floodplain along the stream. This floodplain map is useful for regulation, planning, and management of storm-water-related issues.

Floodplain mapping by the conventional method normally requires mapping the floodplain manually, which can be a very tedious process. It is done by transferring the water surface profiles to the study basemap (usually a contour map) manually. Water surface elevations are mapped at the cross sections and interpolated between cross sections. The mapping is achieved with engineering judgment to follow contours between cross sections on the basemap. Hand-drawn floodplain boundaries are digitized in a CAD program and plotted out as the desired floodplain map.

Instead of manually mapping the floodplain boundaries, HEC-RAS can export the computed water surface profile in a file format that is directly readable by GIS StreamPro. GIS StreamPro processes the water surface profile with the original TIN of the ground surface to obtain the floodplain. The floodplain will honor all the topographic features defined by the TIN; no judgment is involved in this process.

The shapefile of the floodplain created by GIS StreamPro can be imported to a CAD program (if necessary) and overlaid with the study basemap. The floodplain boundaries created by GIS StreamPro may appear to be jagged. If this is the case, the size of the computation grid used for the floodplain mapping may be reduced. This is a minor adjustment in GIS StreamPro; no additional HEC-RAS analysis is required.

COMPARISON OF EFFICIENCY

The most time-consuming steps of the conventional method are:

- Setting up the cross sections (especially if the cross-section coordinates must be scaled from a topographic map)
- Measuring the reach lengths between adjacent cross sections
- Manually entering all of the cross-section data into HEC-RAS
- Manually mapping the floodplain boundaries

The most time-consuming steps of using GIS StreamPro are:

- Creating the TIN model
- Drawing the cross sections, reach lengths, centerline, and bank lines

DESCRIPTION OF EFFICIENCY CASE STUDY

The case study focuses on a reach of Cedar Creek Tributary number 1 in Lufkin, Texas. The reach of the natural stream is 4,000 feet long. Ten channel sections are field surveyed along the reach at intervals of about 500 feet. The reach includes one culvert roadway crossing. Aerial mapping was used to define the general topography of the watershed and 2-foot contours were created from the aerial mapping.

In order to compare the efficiency of the conventional method with the GIS StreamPro method, two tasks were performed, to produce the 100-year floodplain map for the reach using the conventional method and to produce the same floodplain map using GIS Stream-Pro. During the process of each task, time consumed for major steps was recorded.

The conventional method uses the surveyed channel section and the overbank data obtained from the contours to create the cross sections. Reach lengths are measured from basemaps by hand. All the data are input to HEC-RAS manually. Floodplain mapping is first done manually on a basemap. The final floodplain map is produced after digitizing the boundaries into AutoCAD.

GIS StreamPro used the same survey points obtained for the conventional method as part of the input data for the creation of a TIN. The 2-foot contours created from aerial mapping are converted into three-dimensional coordinate points. These are then combined with the field-surveyed points to create the TIN. The cross sections, flow paths, centerline, and bank lines are drawn on the study basemap using a CAD program. This CAD drawing is then exported to ArcView GIS, where it is used by GIS StreamPro.

Table 1 and figure 1 compare the results for each method.

TABLE 1. RESULTS OF EFFICIENCY CASE STUDY

General procedure	Conventional method steps	Time (min.)	GIS StreamPro steps	Time (min.)
Data collection and preparation	Planning field survey for channel data	N/A	Planning field survey for channel data	N/A
	Obtain overbank data from topographic maps	120	Obtain terrain data	10
	Organize and merge surveyed channel data with overbank data to create full cross sections	120	Merge terrain data with survey data and create TIN in GIS StreamPro	30
	Prepare basemap and measure reach lengths	30	Draw cross sections, flow paths, and bank lines	45
			Process and export data for HEC-RAS	5
	Subtotal	**270**		**90**
Model setup and execution	Input data to HEC-RAS	75	Import data to HEC-RAS	5
	Complete the HEC-RAS	30	Complete the HEC-RAS	30
	Execute model, evaluate results, and fine-tune the model as necessary	30	Execute model, evaluate results, and fine-tune the model as necessary	30
			Export floodplain data for GIS StreamPro	5
	Subtotal	**135**		**70**
Floodplain mapping	Map floodplain manually	30	Import floodplain data to StreamPro and process to obtain floodplain	5
	Digitize floodplain to produce the final floodplain map	30	Import floodplain into CAD and fine-tune the floodplain	30
	Subtotal	**60**		**35**
Total		**465**		**195**

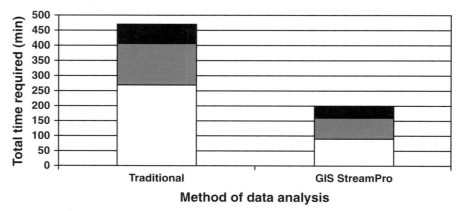

Figure 1. Comparison of times required for floodplain analysis.

DISCUSSION OF EFFICIENCY CASE STUDY RESULTS

The case study shows that GIS StreamPro is much more efficient than conventional methods in performing two general procedures of floodplain analysis. GIS StreamPro required only about one-third of the data collection and preparation. The obvious reason for the time efficiency is that the GIS StreamPro method eliminates much of the need for manual manipulation of data, which is reflected in three key areas: organizing and preparing cross sections, manual input of data in HEC-RAS, and floodplain mapping.

If any of the data preparation and processing steps must be redone (as is commonly the case, because of minor adjustments in some item of input data), the advantages of GIS StreamPro are even more dramatic. The existing TIN could be revised (assuming that any needed changes are relatively minor) and could then be used to quickly revise all cross-section coordinates (using the cross-section alignments and other data still available from the earlier work). This can produce significant savings if a new cross section must be inserted between two existing cross sections, for example.

When a conventional analysis is performed, the floodplain must be mapped only once if possible, so the floodplain mapping is typically put off until the very end of the project, so that there can be no more changes to the computed water surface elevations. If there were such a change after the floodplain had already been mapped, the entire floodplain mapping might have to be done again.

In comparison, GIS StreamPro can automatically compute and plot the floodplain boundaries at any step in the analysis. This means that preliminary floodplain boundaries are available for review much earlier in the study period.

In conclusion, GIS StreamPro and HEC-RAS can increase the efficiency of performing a floodplain analysis tremendously when compared to the conventional method. With more powerful computers, better terrain data, and more data and drawings becoming available in digital format, GIS StreamPro has the potential to become even more efficient in floodplain analysis.

ACCURACY OF COMPUTED WATER SURFACE PROFILES

Since water surface profile computations are used to determine floodplain elevations and boundaries, computed water surface profiles need to be accurate. The Hydrologic Engineering Center of the U.S. Army Corps of Engineers published the results of a study of the accuracy of computed water surface profiles in 1986 (USACE, 1986). The 1986 HEC study, along with later work done at the Waterways Experiment Station (Freeman, Copeland, and Cowan, 1996) provides the basis for determining the most important sources of errors in the computation of water surface profiles using a one-dimensional, gradually varied program such as HEC-RAS. The results of these studies show the following:

- The single most important source of error in computing water surface profiles is the subjective selection of loss parameters, particularly the Manning roughness coefficient, or "n" value.

- The availability of a sufficient amount of cross-section data was also a crucial factor in the accuracy of computed water surface profiles.

The automated floodplain software considered in this analysis does not provide any assistance with the selection of a Manning roughness coefficient. However, it can be very useful in the preparation of adequate cross-section data.

Errors in topographic data for water surface profiles

There is currently no source of topographic data that provides higher accuracy than traditional field survey methods using modern equipment. When the location, orientation, and length of cross sections are planned in advance by an experienced hydraulic engineer, and then carried out by an experienced and professional field survey crew, the resulting information is almost ideal for performing an accurate water surface profile computation.

Too often, however, budget and schedule restrictions mean that fewer cross sections may be obtained. This leads to excessive reach lengths, which can cause several different warnings and error messages during the computation of the water surface profile. The distance between cross sections is too long if hydraulic properties of the flow change too radically from cross section to cross section. If, from one cross section to the next, the slope of the energy grade line decreases by more than 50% or increases by more than 100%, the reach length may be too long for accurate determination of energy losses caused by boundary friction. For example, figure 2 illustrates a HEC-RAS plot of energy slope versus channel station. Each point on the chart represents the energy slope computed at a particular cross section along the stream channel. Note that there are three cross-section locations (including the starting cross section) at which there are "spikes" in the computed energy grade line elevations. Inserting additional cross sections reduces one of these "spikes" considerably, as the figure also indicates.

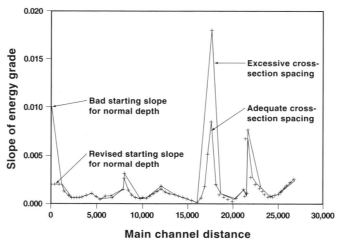

Figure 2. Example of excessive cross-section spacing.

Excessive change in the computed velocity head is another indication of excessive cross-section spacing.

Increasing the accuracy of topographic data

Since the HEC research (USACE, 1986) indicates that the biggest source of uncertainty in the use of topographic data for water surface profile computation is an inadequate number of cross sections, it is logical to conclude that one of the best ways to reduce the uncertainty of a computed water surface profile is to obtain more cross sections. This is true even if the data points that make up each of the new cross sections are less accurate.

Figure 3 illustrates two computed water surface profiles for the same stream channel. One profile (the solid line) was computed with cross sections made up of accurate data points that were spaced too far apart. The excessive reach lengths caused the HEC-RAS program to overshoot the water surface profile. When more cross sections were added, the profile computation became much more accurate, in spite of the fact that the data points used for the additional cross sections

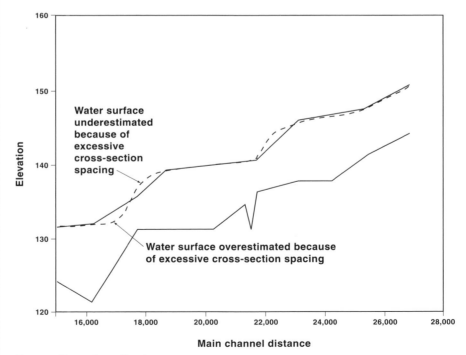

Figure 3. Example profile plot.

were taken from relatively low-quality sources of topographic data. The dashed line shows the more accurate water surface profile.

Having an adequate number of cross sections has some implications for the selection of input data sources and data preparation methods for a HEC-RAS analysis. The ideal data source would provide the following characteristics:

- The data set would provide a dense coverage of elevation values along the length of the stream channel and floodplain, so that cross sections could be laid out in the office, or additional cross sections could be easily added, to ensure that reach lengths were not excessive.

- The individual data points would meet established accuracy standards suitable for water surface profile computations (field survey accuracy is normally not required except in the channel portion of a cross section).

A combination of traditional field survey methods (for channel cross sections and hydraulic structures) and remote sensing (either traditional aerial photogrammetry or LIDAR) for floodplain areas meets these requirements very well.

The ideal data entry method would provide these characteristics:

- Cross sections could be laid out at any location and along any alignment desired for the hydraulic analysis.

- Additional cross sections could be inserted at any time, to fill in excessive reach lengths.

- The full accuracy and detail of the data along the selected cross-section alignments would be available for the hydraulic calculations.

A combination of GIS StreamPro software, when used with ArcView GIS, ArcView 3D Analyst, and HEC-RAS, meets these requirements very well.

HUMAN ERRORS IN DATA ENTRY

In addition to providing an adequate number of cross sections for the analysis, the possibility of human error during data preparation and data entry must be considered.

From human error research in general, we know that error is present in all cognitive processes (Baars, 1992; Reason, 1990). Human cognitive processes are very fast and flexible, but their methods of operation inherently produce a small error rate.

Research in other human cognitive activities indicates that errors are not merely inevitable; they are actually predictable. Research has produced human error data from a large number of experiments and real-world incidents across many types of human cognitive activity. These data collectively suggest that human beings have a natural uncorrected error rate that ranges from about 1 error per 1,000 operations (0.1%) up to about 1 error per 10 operations (10%), with an average of about 1 error per 100 operations (1%). Of course the error rate will depend on the task, as will the error detection and correction rate. The level of experience of the person performing the task has some influence on the rate of errors, but even an experienced person makes errors. Research indicates that the error rates of novices and experts are different, but not dramatically so.

Therefore, the only question is the rate of uncorrected errors, that is, errors that are not detected and fixed by people as they work. Studies in statistics problem solving (Allwood, 1984) and writing (Hayes & Flower, 1980) indicate that people both detect and correct errors as they go along. People also occasionally perform more systematic error checking, in which they go back over their work before they finish. They also miss many of their errors. For simple slips, the detection and correction rate is over 90%. For more complex errors, the detection and correction rate is much lower. Experts do not have a dramatically higher rate of error detection and correction than do novices (Galletta et al., 1993; Grudin, 1983).

Computer programmers have to spend about a third of the total development time on systematic error checking after development. Otherwise, programs will have an unacceptably high number of errors upon delivery. Often, this systematic error checking takes the form of team code inspection (Fagan, 1976) because both experiments and field experience have shown that individual inspectors will not catch a large portion of all programming errors (Basili and Selby, 1986; Myers, 1976).

TECHNIQUE FOR HUMAN ERROR PREDICTION (THERP)

Therefore, we should expect that human beings processing and entering a large amount of data for a floodplain analysis will make some mistakes. How many mistakes? This section will describe a method of providing an estimate.

The study of human errors is called Human Reliability Analysis (HRA), and one of the most popular quantitative methods of HRA is the Technique for Human Error Rate Prediction (THERP). THERP was originally developed at the Sandia Laboratories in the 1950s as a way of analyzing the risks involved in the assembly of nuclear weapons (Swain and Guttman, 1983). THERP provides a mechanism for modeling as well as quantifying human error. THERP involves the following steps:

1. Perform a task analysis to identify all of the individual tasks of the process. Perform a walk-through or talk-through of the tasks, on site if possible, with as much input from process participants and systems analysts as possible.

2. Develop HRA event trees for the procedure. The HRA event trees provide a graphical representation of the task analysis.

3. Assign nominal human error probabilities (HEPs). HEPs are available from standard databases (e.g., tables 5-8 to 5-37 in Gertman and Blackman, 1994), or can be determined by empirical study. Expert judgments can be used if necessary.

4. Estimate the relative effects of performance-shaping factors (PSFs). PSFs include environmental factors (e.g., temperature, humidity), shift rotation, organizational structure, reward structure, instructions, perceptual and motor requirements of the task, task criticality, frequency, repetitiveness, psychological stressors (like suddenness of onset, task load, and high risk), and internal factors (such as previous training, personality, and motivation).

5. Assess the degree of dependence between any two actions.

6. Determine success and failure probabilities.

7. Determine the effects of error recovery factors.

THERP assumes that particular types of errors, such as reading or omitting an instructional step, are presumed to occur at constant rates. If the tasks that a person performs can be broken down into subtasks for which these types of errors can be predicted, then the probability of the successful completion of the overall task can be predicted.

The database of error rates used by THERP has evolved over a 30-year period, using a combination of statistical data and expert judgment, and is presumed to be accurate within an order of magnitude. When studies have been conducted to verify this assumption, estimates are usually found to be much closer, varying by factors of 2 or 3 rather than 10.

A complete THERP study is very comprehensive, and is not required for this analysis. However, a preliminary study of the potential for error in manual data entry of cross-section data could provide useful information.

Three basic operations may be required for HEC-RAS data entry. According to available published data, each of these three operations have dramatically different human error rates:

- Manual data entry from a graphical source (such as a contour map). This is actually a fairly complex operation, because it requires that a value be scaled or measured from a map.

- Manual data entry from a printed or tabulated source, such as reduced survey field notes.

- Direct data transfer from the data source to the program performing the analysis.

The error rates for each of these operations may be estimated from available HRA authorities. For example, table 5-17 (page 134) of Gertman and Blackman, 1994, lists several measured error rates for reading and recording values from analog meters, digital meters, chart recorders, and graphs. These error rates range from 0.1% for digital meters up to 5% for a printing recorder with a large number of parameters. A contour map is assumed to be equivalent to a graph, which has an estimated error rate of 1%, whereas a printed or tabulated data source is assumed to be equivalent to a digital meter, which has an error rate of 0.1%. Direct data transfer is assumed to have a data entry error rate near zero.

In spite of the low rate of error at each step, however, the probability of error increases tremendously because of the large number of input data values required for most cross sections. As figure 4 indicates, the probability of human error in manually entering the data for one cross section is between 2% and 10%, assuming that the cross section includes about 100 ground coordinates. For the maximum cross-section size currently supported by HEC-RAS (500 points), the expected probability of errors in a cross section would rise to the range of 5% to 50%, depending on the data source.

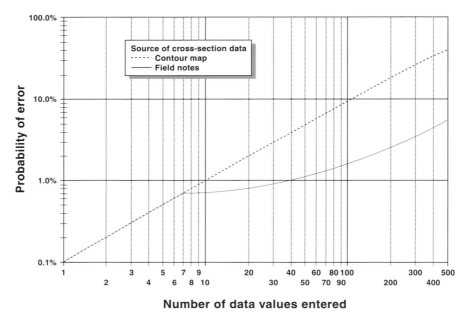

Number of data values entered

Figure 4. Probability of human error during cross-section data entry.

CONCLUSIONS

This paper has presented a comparison of two methods of performing floodplain studies: a traditional method that involves manual data entry, and an automated method that uses GIS StreamPro, a software product distributed by Dodson & Associates, Inc., and designed to work seamlessly with ArcView GIS, ArcView 3D Analyst, and the U.S. Army Corps of Engineers HEC-RAS River Analysis System.

A case study showed that GIS StreamPro eliminates about two-thirds of the effort required to perform a floodplain study, even though the use of GIS StreamPro requires the preparation of a TIN model for the study area. The primary efficiency advantages of GIS StreamPro are in the elimination of time-consuming steps such as data entry and manual floodplain mapping.

GIS StreamPro provides another significant advantage: recomputing floodplain elevations and remapping floodplain boundaries is fully automated and can be redone almost immediately.

An analysis of accuracy considerations showed that, without GIS StreamPro, human errors may be expected in 1% to 10% of all normal cross sections (and more frequently in longer cross sections). However, since GIS StreamPro provides for direct data transfer to HEC-RAS, human errors are practically eliminated.

GIS StreamPro provides another advantage: inserting additional cross sections to eliminate excessive reach lengths is very easy. The ability to quickly and easily insert additional cross sections addresses one of the major sources of error in computed water surface profiles: excessive cross-section spacing due to poor initial cross-section layout or an inadequate number of available cross sections.

REFERENCES

Baars, B. J. (ed.) 1992. *Experimental Slips and Human Error*. New York: Plenum.

Basili, V. R., and R. W. Selby, Jr. 1986. Four Applications of a Software Data Collection and Analysis Methodology. In *Software System Design Methods,* ed. J. K. Skwirzynski. Berlin: Springer–Verlag.

Fagan, M. E. 1976. Design and Code Inspections to Reduce Errors in Program Development. *IBM Systems Journal* 15(3):182–211.

Freeman, G. E., R. R. Copeland, and M. A. Cowan. 1996. Uncertainty in Stage-Discharge Relationships. *Proceedings, Seventh IAHR International Symposium on Stochastic Hydraulics,* Mackay, Queensland, Australia, IAHR. In publication.

Galletta, D. F., D. Abraham, M. El Louadi, W. Lekse, Y. A. Pollalis, and J. L. Sampler. An Empirical Study of Spreadsheet Error-Finding Performance. *Accounting Management Information Technology* 3(2).

Gertman, David I., and Harold S. Blackman. 1994. *Human Reliability and Safety Analysis Data Handbook*. New York: John Wiley & Sons, Inc.

Grudin, J. T. 1983. Error Patterns in Novice and Skilled Transcription Typing. Chapter 6 in *Cognitive Aspects of Skilled Typewriting,* ed. W. E. Cooper, 121–43. New York: Springer–Verlag.

Hayes, J. R., and L. S. Flower. 1980. Identifying the Organization of Writing Processes. In *Cognitive Processes in Writing,* eds. L. Gregg and E. Steinberg, 3–30. Hillsdale, New Jersey: Erlbaum.

Myers, G. J. 1978. A Controlled Experiment in Program Testing and Code Walk-throughs/Inspections. *Communications of the ACM* 21(9) 760–68.

Reason, J. 1990. *Human Error*. Cambridge, United Kingdom: Cambridge University Press.

Swain, A. D., and H. E. Guttman. 1980. *Handbook of Human Reliability Analysis with Emphasis on Nuclear Power Plant Applications*. (Draft report for interim use and comment.) Technical Report NUREG/CR-1278, U.S. Nuclear Regulatory Commission, Washington, D.C., October 1980. Cited in B. Kantowitz and R. D. Rorkin, 1983, *Human Factors: Understanding People–System Relationships*, New York: John Wiley & Sons, Inc.

U.S. Army Corps of Engineers (USACE). 1986. Accuracy of Computed Water Surface Profiles. Research Document 26. Hydrologic Engineering Center, Davis, California.

U.S. Army Corps of Engineers (USACE). 1998. *HEC-RAS River Analysis System Version 2.2 Users Manual*. Hydrologic Engineering Center, Davis, California.

ABOUT THE AUTHORS

Roy D. Dodson, P.E. is president of Dodson & Associates, Inc., in Houston, Texas, a firm that he founded in 1983. Mr. Dodson is recognized as a national authority on floodplain management, hydrology and hydraulics, and storm water pollution control, having written numerous books, articles, and professional papers on these topics. He has presented more than two hundred training seminars to engineers in these fields over the past 15 years in locations throughout the United States. His firm is engaged in drainage planning for projects in the Houston area and in many other locations. He has also performed floodplain and environmental consulting work for many major U.S. corporations and for several agencies of the federal government.

Xiaojian Li joined the staff at Dodson & Associates, Inc., in 1997 shortly after receiving his bachelor of science degree in civil engineering at Rice University. During his time with the firm, Mr. Li contributed significantly to many important projects. He is currently attending Stanford University's Civil Engineering Masters program, and plans to return to Dodson & Associates, Inc., upon his graduation.

CONTACT THE AUTHORS

Roy D. Dodson and Xiaojian Li
Dodson & Associates, Inc.
5629 FM 1960 West, Suite 314
Houston, TX 77069-4216
Telephone: 1-800-235-8069 (U.S. or Canada), 281-440-3787
Fax: 281-440-4742
software@dodson-hydro.com
www.dodson-hydro.com

Selected Bibliography

FURTHER READING

GIS applications in water resources is a dynamic field and a constant stream of new publications follows the latest developments. Finding a good pathway into this range of literature depends on the depth and focus of needed information. Some pathways enter from the GIS side of the field, while others enter from the water resources side. This section presents some of the available sources from both realms.

A good starting point is the GIS Bibliography, housed in ESRI's Virtual Campus Library on the Web at campus.esri.com. The GIS Bibliography contains references to thousands of GIS-related papers, articles, and technical reports. Many of the items can be read online in their full-text versions. The GIS Bibliography continues to grow as a public, free, annotated source of information on the GIS field.

PERIODICALS

The following GIS journals often cover applications in water resources:

International Journal of Geographical Information Science, published by Taylor & Francis, is searchable at www.tandfdc.com/JNLS/gis.htm.

Photogrammetric Engineering and Remote Sensing, the flagship journal of the American Society for Photogrammetry and Remote Sensing, is searchable at www.asprs.org.

Surveying and Land Information Systems. The last few years of this journal from the American Congress on Surveying and Mapping (ACSM) can be searched at www.survmap.org/commun42/salis01.htm.

Many water resources journals provide peer-reviewed articles on GIS applications. Some have special issues on GIS applications in various fields of water resources.

Water Resources Research, published by the American Geophysical Union, is an interdisciplinary journal integrating research in the social and natural sciences of water. The AGU's Web site (www.agu.org/wrr) includes information about the journal.

The American Society of Civil Engineers produces several relevant journals that can be searched on ASCE's Web site (www.pubs.asce.org):

> *Journal of Hydrologic Engineering*
>
> *Journal of Computing in Civil Engineering*
>
> *Journal of Hydraulic Engineering*
>
> *Journal of Water Resources Planning and Management*

Hydrological Processes and Earth Surface Processes and Landforms from John Wiley & Sons can be searched from the Wiley Interscience Web site (www.interscience.wiley.com).

Journal of Hydrology from Elsevier Science (www.elsevier.com) produced a special issue on hydrological applications of GIS in May 1998 (vol. 12, no. 5) and has frequent articles on GIS applications.

Journal of the American Water Resources Association (formerly *Water Resources Bulletin*). You can look through tables of contents for the last several years of the journal at AWRA's Web site (www.awra.org/jawra/index.html).

CONFERENCE PROCEEDINGS

Regularly published GIS conferences that include significant information on water resources include the following:

Proceedings of the ESRI International User Conference. Held since 1985, this event is the largest annual conference on GIS. All conference papers since 1995 are available at ESRI's Virtual Campus Library (campus.esri.com) in the GIS Bibliography; most include full-text papers. The same papers can also be found at ESRI's corporate Web site (www.esri.com).

Conference/Workshop on Integrating GIS and Environmental Modeling. The National Center for Geographic Information and Analysis (NCGIA) has held three of these conferences so far (1991, 1993, and 1996), with a fourth scheduled for 2000. More information can be found at www.ncgia.org.

The proceedings of the first two conferences were published as edited books:

Goodchild, Michael, Bradley Parks, and Louis Steyaert. *Environmental Modeling with GIS*. New York, New York: Oxford University Press, 1993.

Goodchild, Michael, Louis Steyaert, and Bradley Parks. *GIS and Environmental Modeling*. Fort Collins, Colorado: *GIS World*, 1996.

ACSM/ASPRS Annual Convention and Exposition. From the mid-1980s to the late 1990s, the American Congress on Surveying and Mapping (ACSM) and the American Society for Photogrammetry and Remote Sensing (ASPRS) held a joint annual conference that consistently produced papers on water resources and GIS. The two organizations now hold separate conferences, which remain good sources of papers. More information is on their Web sites: ACSM at www.survmap.org and ASPRS at www.asprs.org.

GIS/LIS Annual Conference and Exposition Proceedings. Sponsored by six leading GIS professional organizations, the GIS/LIS conference took place annually from 1988 to 1998. The last two years are available through the GIS Bibliography at campus.esri.com or on CD–ROM from the Urban and Regional Information Systems Association (URISA) or Association of American Geographers (AAG). Previous years were published in book form and some years are still available from AAG, URISA, ACSM, or the Geospatial Information and Technology Association (GITA).

International Symposium on Computer-Assisted Cartography. Usually known as Auto-Carto or Autocarto, this conference coincided with the ACSM/ASPRS Annual Convention and Exposition throughout the 1990s. In fact, the proceedings were sometimes included as a volume of the ACSM/ASPRS proceedings, though they occasionally found their way into a monograph issue of the *Cartographica* journal or were published by other means.

In the water resources area, the American Society of Civil Engineers, the American Water Resources Association, and the International Association of Hydrological Sciences hold conferences that consistently cover GIS implementation in the field. Often, a specialty symposium on GIS and water resources is held in conjunction with their annual conferences.

Geographic Information Systems and Water Resources. The American Water Resources Association (AWRA) held its first symposium on Geographic Information Systems and Water Resources in 1993 in Mobile, Alabama. A second symposium took place in 1996 in Fort Lauderdale, Florida. The proceedings of these meetings include papers on database issues, water quality, groundwater, storm water, and watershed studies. AWRA's Web site (www.awra.org) has more information.

Remote Sensing and Geographic Information Systems for Design and Operation of Water Resources Systems. Held during the Fifth Scientific Assembly of the International Association of Hydrological Sciences (IAHS) at Rabat, Morocco, 1997, this international symposium was sponsored by the IAHS, the United Nations Educational, Scientific and Cultural Organization, the World Meteorological Organization, and the Moroccan Water Resources Association.

Application of Geographic Information Systems in Hydrology. Vienna, Austria, was the site of two conferences, in 1993 and 1996. Known as HydroGIS, these events were organized and published by the International Association of Hydrological Sciences.

Other books from ESRI Press

ESRI Special Editions

GIS for Everyone
Now everyone can create smart maps for school, work, home, or community action using a personal computer. Includes the ArcExplorer™ geographic data viewer and more than 500 megabytes of geographic data. ISBN 1-879102-49-8

The ESRI Guide to GIS Analysis
By the author of the best-selling GIS classic *Zeroing In: GIS at Work in the Community* comes an important new book about how to do real analysis with a geographic information system. *The ESRI Guide to GIS Analysis, Volume 1: Geographic Patterns and Relationships* focuses on six of the most common geographic analysis tasks. ISBN 1-879102-06-4

Modeling Our World
With this comprehensive guide and reference to GIS data modeling and to the new geodatabase model introduced with ArcInfo 8, you'll learn how to make the right decisions about modeling data, from database design and data capture to spatial analysis and visual presentation. ISBN 1-879102-62-5

ESRI Map Book: Implementing Concepts of Geography
A full-color collection of some of the finest maps produced using GIS software. ISBN 1-879102-60-9

Hydrologic and Hydraulic Modeling Support with Geographic Information Systems
This book presents the invited papers in Water Resources at the 1999 ESRI International User Conference. Covering practical issues related to hydrologic and hydraulic water quantity modeling support using GIS, the concepts and techniques apply to any hydrologic and hydraulic model requiring spatial data or spatial visualization. ISBN 1-879102-80-3

The Case Studies Series

ArcView GIS Means Business
Written for business professionals, this book is a behind-the-scenes look at how some of America's most successful companies have used desktop GIS technology. The book is loaded with full-color illustrations and comes with a trial copy of ArcView GIS software and a GIS tutorial. ISBN 1-879102-51-X

Zeroing In: Geographic Information Systems at Work in the Community
In twelve "tales from the digital map age," this book shows how people use GIS in their daily jobs. An accessible and engaging introduction to GIS for anyone who deals with geographic information. ISBN 1-879102-50-1

Serving Maps on the Internet
Take an insider's look at how today's forward-thinking organizations distribute map-based information via the Internet. Case studies cover a range of applications for ArcView Internet Map Server technology from ESRI. This book should interest anyone who wants to publish geospatial data on the World Wide Web. ISBN 1-879102-52-8

Managing Natural Resources with GIS
Find out how GIS technology helps people design solutions to such pressing challenges as wildfires, urban blight, air and water degradation, species endangerment, disaster mitigation, coastline erosion, and public education. The experiences of public and private organizations provide real-world examples. ISBN 1-879102-53-6

Enterprise GIS for Energy Companies
A volume of case studies showing how electric and gas utilities use geographic information systems to manage their facilities more cost effectively, find new market opportunities, and better serve their customers. ISBN 1-879102-48-X

MORE ESRI PRESS TITLES ARE LISTED ON THE NEXT PAGE ➤

ESRI educational products cover topics related to geographic information science, GIS applications, and ESRI technology. You can choose among instructor-led courses, Web-based courses, and self-study workbooks to find education solutions that fit your learning style and pocketbook. Visit www.esri.com/education for more information.

ESRI Press ■ 380 New York Street ■ Redlands, California 92373-8100

Other books from ESRI Press

The Case Studies Series CONTINUED

Transportation GIS
From monitoring rail systems and airplane noise levels, to making bus routes more efficient and improving roads, this book describes how geographic information systems have emerged as the tool of choice for transportation planners. ISBN 1-879102-47-1

GIS for Landscape Architects
From Karen Hanna, noted landscape architect and GIS pioneer, comes *GIS for Landscape Architects*. Through actual examples, you'll learn how landscape architects, land planners, and designers now rely on GIS to create visual frameworks within which spatial data and information are gathered, interpreted, manipulated, and shared. ISBN 1-879102-64-1

GIS for Health Organizations
Health management is a rapidly developing field, where even slight shifts in policy affect the health care we receive. In this book, you'll see how physicians, public health officials, insurance providers, hospitals, epidemiologists, researchers, and HMO executives use GIS to focus resources to meet the needs of those in their care. ISBN 1-879102-65-X

ESRI Software Workbooks

Understanding GIS: The ARC/INFO Method (UNIX/Windows NT version)
A hands-on introduction to geographic information system technology. Designed primarily for beginners, this classic text guides readers through a complete GIS project in ten easy-to-follow lessons. ISBN 1-879102-01-3

Understanding GIS: The ARC/INFO Method (PC version)
ISBN 1-879102-00-5

ARC Macro Language: Developing ARC/INFO Menus and Macros with AML
ARC Macro Language (AML) software gives you the power to tailor workstation ARC/INFO software's geoprocessing operations to specific applications. This workbook teaches AML in the context of accomplishing practical workstation ARC/INFO tasks, and presents both basic and advanced techniques. ISBN 1-879102-18-8

Getting to Know ArcView GIS
A colorful, nontechnical introduction to GIS technology and ArcView GIS software, this workbook comes with a working ArcView GIS demonstration copy. Follow the book's scenario-based exercises or work through them using the CD and learn how to do your own ArcView GIS project. ISBN 1-879102-46-3

Extending ArcView GIS
This sequel to the award-winning *Getting to Know ArcView GIS* is written for those who understand basic GIS concepts and are ready to extend the analytical power of the core ArcView GIS software. The book consists of short conceptual overviews followed by detailed exercises framed in the context of real problems. ISBN 1-879102-05-6

ESRI Press publishes a growing list of GIS-related books. Ask for these books at your local bookstore or order by calling 1-800-447-9778. You can also shop online at www.esri.com/gisstore. Outside the United States, contact your local ESRI distributor.

ESRI Press ■ 380 New York Street ■ Redlands, California 92373-8100